We Speak Chinese

我们说中文

课本 初级 1

Textbook Beginner

宋可音（Ko-Yin Sung）编著

北京大学出版社
PEKING UNIVERSITY PRESS

图书在版编目(CIP)数据

我们说中文.初级1 / 宋可音 编著.—北京：北京大学出版社，2013.12
ISBN 978-7-301-23443-3

Ⅰ.我… Ⅱ.宋… Ⅲ.汉语—对外汉语教学—教材 Ⅳ.H195.4

中国版本图书馆CIP数据核字(2013)第266188号

书　　名：我们说中文·初级1
著作责任者：宋可音　编著
责任编辑：沈　岚
标准书号：ISBN 978-7-301-23443-3/H·3426
出版发行：北京大学出版社
地　　址：北京市海淀区成府路205号　100871
网　　址：http://www.pup.cn　　新浪微博：@北京大学出版社
电子信箱：zpup@pup.pku.edu.cn
电　　话：邮购部 62752015　发行部 62750672　编辑部 62767349　出版部 62754962
印 刷 者：北京大学印刷厂
经 销 者：新华书店
889毫米×1194毫米　大16开本　12.5印张　380千字
2013年12月第1版　2013年12月第1次印刷
定　　价：75.00元(课本、练习册全2册,含1张MP3光盘)

目录
Contents

前言

介绍

本系列教材分为两级，共有四册课本和练习册。

课本	练习册
初级1	初级1
初级2	初级2
中级1	中级1
中级2	中级2

课本和练习册配合使用。本系列教材主要是为汉语为第二语言的初中生和高中生设计的。当然，这套教材也可以作为大学一年级汉语课程的补充教材或用于大学的汉语会话及文化课程。

初级1和初级2适用于第一年学习汉语的中学生。中级1和中级2则适用于第二年学习汉语的中学生。教师可根据课程时间长短和学习者的能力加以选择。

本系列教材的设计理念以培养语言交际能力作为重点，教学以沟通的有效性为主要目的，而不是语法结构的讲解，旨在帮助学习者以口语和书面语的形式用目标语中文与他人沟通，主题的选择是调查了约100名美国中学生最喜欢的中文教材后决定的。

课本的结构

初级1第一课主要介绍汉语拼音，第二课主要介绍汉字系统。其余课基本包括以下几个部分：

沟通任务：列出学习者在学完本课后能应用在日常生活中的功能点。

课前讨论：课前讨论列出的问题能让老师知道学习者对每课主题的了解程度。这些问题可以帮助老师决定花多少时间为学习者建立主题的背景知识。虽然列出的问题是英文，如果学习者对每课的主题准备充分，老师也可鼓励学习者试着用目标语中文来沟通。

生词：这一部分列出课文里使用的词语（包括专有名词）。简体和繁体汉字一并列出。另外，词语的拼音、词性以及例句也包括在生词表中。

补充词语：这个部分列出口语沟通活动和读写沟通活动里所使用的词语。为了让学生增加词汇量，此部分的词汇量没有有意控制。补充词语学多学少由老师按照学生的程度决定。

沟通任务：这一部分列出课文情景的主要语言点。这一部分不用传统语法的教学方式来讲解。此部分介绍每个沟通任务及例句，并请学习者与同学一起进行沟通练习。

情景：情景包含每课的课文内容。每个情景展示之前所介绍的词汇和沟通任务。并引导学生学习如何在不同的情景下进行有效的会话。此部分显示的不同情景有助于培养学习者对目标语中文的语感。

口语沟通活动：此部分提供学习者尝试运用刚学到的语言知识练习口头表达能力的机会，以交际语言教学法为原则，设计了不同情景让学习者用目标语中文做有意义的会话练习。与强调演练或记忆词汇的活动相比，本系列教材的沟通活动可以更有效地帮助学习者提高语言交际能力。

读写沟通活动：此部分与口语沟通活动采用相同的设计理念，只是着重于学习者阅读和写作技能的训练。

讨论：此部分介绍有关的中国文化。讨论的话题与每课的主题相关，列出的讨论问题可使老师与学生进一步理解中国文化。

练习册的结构

初级1第一课为汉语拼音练习，第二课为汉字知识的练习。其余课基本包括以下几个部分：听力练习、口语练习、阅读练习、写作练习和沟通练习。前四个练习中的每一个练习分别侧重于听、说、读、写这四种语言技能中的一种，最后一个练习是这四种语言技能的综合练习。

给教师的话

使用本系列教材的教师应注重第二语言通过沟通任务的习得。换句话说，学生在类似于现实生活情景的课堂活动中（例如：小组讨论、模拟和角色扮演）学习汉语。这也是本系列教材所倡导的教学法。当学习者做练习时，教师应支持和帮助学生构建目标语中文的语言知识，从而使他们逐渐成为独立的学习者。

致谢

首先感谢北京大学出版社出版《我们说中文》这一系列教材，使我有机会实践我对汉语教学的想法，同时感谢王飙主任的专业指导和支持。另外，我也要感谢丈夫罗蓝登、母亲曾淑玲和父亲宋总男对我的鼓励和支持！

编者

Preface

Introduction

This book series consists of four textbooks and workbooks in two levels:

Textbook	Workbook
Beginner 1	Beginner 1
Beginner 2	Beginner 2
Intermediate 1	Intermediate 1
Intermediate 2	Intermediate 2

Each textbook is combined with its workbook into a single book. This series is designed mainly for middle and high school students who study Chinese as a second or foreign language. The beginner books can be used as supplementary materials for first-year Chinese at the collegiate level. Also, the intermediate books can be used in a college Chinese conversation or culture course.

In a middle or high school, the beginner books are suitable for use in first-year Chinese while the intermediate books are suitable for use in second-year Chinese. However, teachers need to make their own judgment, taking into consideration the class time period, how the books are used, and the ability of learners. The design of this book series uses the communicative language learning and teaching approaches as the guide. The series focuses on the effectiveness of communication rather than the practice of grammatical patterns. The goal is to assist learners in successfully communicating with others in the target language, Mandarin Chinese, in both oral and written forms. The topic of each lesson was carefully selected based on survey results of approximately 100 American middle and high school students' most desirable Chinese textbook topics.

Structure of the Textbook

Except Lesson 1, which introduces the phonetic system, pinyin, and Lesson 2, which discusses the Chinese writing system, each lesson consists of the following sections:

Communicative Tasks: This section identifies structures learners will study in the lesson and be able to apply in daily life after learning the lesson.

Warm-up Discussion Questions: The warm-up questions are used to learn how much background knowledge learners may have in regard to the lesson topic. It is used as a way to connect with learners' own experiences and help instructors decide how much time should be spent building on learners' background knowledge for the lesson. Although the questions are in English, teachers can encourage students to try to use the target language to communicate if they have enough background knowledge and are familiar with the topic.

Vocabulary: The vocabulary chart in the Vocabulary section illustrates the vocabulary words used in the lesson texts in the Scenarios section. The vocabulary words are shown in both simplified and traditional Chinese characters. In addition, the pinyin of the words and examples of how to use the words are included in the chart.

Supplementary Vocabulary: The vocabulary chart in the Supplementary Vocabulary section includes vocabulary words used in the Oral and Literacy Communication Activities sections. The number of vocabulary in each lesson is not purposely controlled. Teachers can decide how to learn the supplementary vocabulary depending on the learners' language level.

Communicative Tasks: This section lists the main structures used in the Scenarios section. The traditional grammar methods are not used to introduce the structures. Instead, the communicative goal of each structure is identified, examples of how to use the structures are given, and learners are asked to try out the structures with classmates.

Scenarios: The Scenarios section contains the main texts of the lesson. Each scenario in the section provides examples of how the structures and vocabulary words introduced in the previous sections can be used in a particular context. A competent speaker is expected to know what to say to others appropriately in any situation. Hence, showing how the structures are used in different scenarios helps develop learners' target language knowledge and communicative competence.

Oral Communication Activities: This section provides learners with opportunities to try out the newly learned language knowledge with a focus on oral skills. Drawing upon the principle of Communicative Language Learning, the oral activities in this section are designed as communicative tasks in which certain scenarios are created in order for students to work collaboratively to interact in the target language with the focus on meaning rather than form.

Compared to activities which emphasize drills or memorizing vocabulary and structures, the communicative activities in this textbook can more effectively help learners use the target language appropriately in real life situations.

Literacy Communication Activities: This section adopts the same activity design principle mentioned in the previous section except that this section focuses on the practice of reading and writing.

Discussion: This section introduces aspects of Chinese culture. The topic varies in each lesson, but is closely related to the lesson topic. The discussion questions listed in this section provide instructors a starting point for a cultural discussion with their learners to help them gain cultural knowledge of the target language.

Structure of the Workbook

Except Lesson 1, which contains exercises of the pinyin system, and Lesson 2, which has exercises of the knowledge of Chinese characters, each lesson consists of the following sections: Listening Exercise, Speaking Exercise, Reading Exercise, Writing Exercise, and Communicative Exercise. Each of the first four exercises focuses on one of the four language skills while the last exercise provides practice of a combination of language skills.

Notes to Teachers

Teachers who use this book series should keep in mind that second language acquisition develops through communicative tasks. In other words, students learn through class activities which are similar to real-life situations, such as group discussions, real-life simulations, and role-play, all of which this book series advocates and provides. When learners carry out the activities, teachers should work as facilitators to support language development while their students move progressively to independent learners.

Acknowledgments

I would like to express my gratitude to Peking University Press for offering me the opportunity of implementing my Chinese language teaching and learning ideas in this textbook series. Many thanks go to the director, Biao Wang, for his professional guidance and continuous support. I would also like to thank my significant other, Brendan, and my parents, Shuling and Tsungnan, who have always encouraged me to pursue my work on this series.

词性缩略语
Abbreviations of Parts of Speech

Abbreviation	Definition
n.	noun
v.	verb
pron.	pronoun
adj.	adjective
adv.	adverb
part.	particle
conj.	conjunction
m.w.	measure word
num.	number
suff.	suffix
prep.	preposition
aux.	auxiliary
int.	interrogative
v.o.	verb object
interj.	interjection
pre.	prefix
p.n.	proper nouns
ono.	onomatcopoetic

第一课
Lesson 1

汉语拼音
The Chinese Phonetic System — Hanyu Pinyin

汉语拼音 Hanyu Pinyin

Hanyu pinyin is the standard phonetic spelling system in China. The Mandarin Chinese word "hanyu" means the Chinese language. The Mandarin Chinese word "pinyin" means phonetics. Hanyu pinyin, hereafter referred to simply as pinyin, which uses 25 of the 26 letters of the Latin alphabet and tone marks, was first introduced by the Chinese government in 1958 as a means to teach standard Chinese language pronunciation. Pinyin is also widely used as an input method to enter Chinese characters into computers.

Initials and Finals

Initials (known as consonants in the English language) and finals (known as vowels in the English language) are the fundamental elements of pinyin. Most Mandarin Chinese syllables are spelled with either:

(1) one initial followed by a single final (e.g. nǐ)

or

(2) one initial followed by a compound final (e.g. nǎo).

There are a few exceptions in which a syllable is without an initial (e.g. ài).

Initials

Pinyin has 21 initials. Listen to and repeat after the audio recordings. Try your best to imitate the pronunciations.

	Unaspirated	Aspirated	Nasals	Fricatives	Lateral and Retroflex
Lip sounds	b	p	m	f	
Tip of tongue	d	t	n		l
Back of tongue	g	k		h	
Blade of tongue	j	q		x	
Curled back of tongue	zh	ch		sh	r
Hiss sounds	z	c		s	

Practice: Initials

Listen to the recordings, repeat after the recordings each time an initial is pronounced, and write down the initials you hear in the order that you hear them.

1 Lip Sounds: b p m f

(1) ________ (2) ________ (3) ________ (4) ________

2 Tip of Tongue: d t n l

(1) ________ (2) ________ (3) ________ (4) ________

3 Back of Tongue: g k h

(1) ________ (2) ________ (3) ________

4 Blade of Tongue: j q x

(1) ________ (2) ________ (3) ________

5 Curled Back of Tongue: zh ch sh r

(1) ________ (2) ________ (3) ________ (4) ________

6 Hiss Sounds: z c s

(1) ________ (2) ________ (3) ________

Finals

Simple Finals

a	o	e	ê
i	u	ü	

Compound Finals

	i	u	ü
Compound Finals (multiple vowels)			
a	ia	ua	
o		uo	
e			
ê	ie		üe
ai		uai	
ei		ui(uei)[1]	
ao	iao		
ou	iu (iou)[2]		
Nasal Finals (ending with -n or -ng)			
an	ian	uan	üan
en	in	un (uen)[3]	ün (üen)[4]
ang	iang	uang	
eng	ing	ueng	
ong	iong		
Retroflex final (ending with -r)			
er			

1 ui is pronounced as uei, but written as ui

2 iu is pronounced as iou, but written as iu

3 un is pronounced as uen, but written as un

4 ün ir pronounced as üen, but written as ün

Practice: Finals

Listen to the recordings, repeat after the recordings each time a final is pronounced, and write down the finals in the order that you hear them.

1 Simple Finals: a o e ê i u ü

(1) _____ (2) _____ (3) _____ (4) _____ (5) _____ (6) _____ (7) _____

2 Compound Finals

2.1 ai ei ao ou

(1) _________ (2) _________ (3) _________ (4) _________

2.2 ia iao ie iu

(1) _________ (2) _________ (3) _________ (4) _________

2.3 ua uo uai ui üe

(1) _________ (2) _________ (3) _________ (4) _________ (5) _________

3 Nasal Finals and Retroflex Final

3.1 an en ang eng ong

(1) _________ (2) _________ (3) _________ (4) _________ (5) _________

3.2 ian in iang ing iong

(1) _________ (2) _________ (3) _________ (4) _________ (5) _________

3.3 uan un(uen) uang ueng

(1) _________ (2) _________ (3) _________ (4) _________

3.4 üan ün(üen) er

(1) ________ (2) ________ (3) ________

Spelling Rules

Rule 1. When spelled with an initial, the final "iou" is spelled "iu" instead.

(Initials)	+	(Final)		
m				miu
d				diu
n				niu
l	+	iou	=	liu
j				jiu
q				qiu
x				xiu

Rule 2. When spelled with an initial, the final "uei" is spelled "ui" instead.

(Initials)	+	(Final)		
d				dui
t				tui
z				zui
c				cui
s				sui
zh	+	uei	=	zhui
ch				chui
sh				shui
r				rui
g				gui
k				kui
h				hui

Rule 3. When spelled with an initial, the final "uen" is spelled "un" instead.

(Initials)	+	(Final)		
d				dun
t				tun
l				lun
z				zun
c				cun
s				sun
zh	+	uen	=	zhun
ch				chun
sh				shun
r				run
g				gun
k				kun
h				hun

Rule 4. When spelled with the initials "j", "q", or "x", the final "ü" is spelled "u" instead.

Initial "j"

(Initial)	+	(Finals)		
j	+	ü	=	ju
j	+	üe	=	jue
j	+	üan	=	juan
j	+	ün	=	jun

Initial "q"

(Initial)	+	(Finals)		
q	+	ü	=	qu
q	+	üe	=	que
q	+	üan	=	quan
q	+	ün	=	qun

Initial "x"

(Initial)	+	(Finals)		
x	+	ü	=	xu
x	+	üe	=	xue
x	+	üan	=	xuan
x	+	ün	=	xun

Rule 5. The following syllables are pronounced the same as their initials.

(Initials	=	Syllables)
z	=	zi
c	=	ci
s	=	si
zh	=	zhi
ch	=	chi
sh	=	shi
ri	=	ri

Rule 6. When the finals "i", "u" and "ü" do not follow any initials, "i" is written as "y", "u" is written as "w", and "ü" is written as "yu".

i → y

i	→	y
i		yi
ia		ya
iao		yao
ie		ye
iou		you
ian	→	yan
in		yin
iang		yang
ing		ying
iong		yong

u → w

u	→	w
u		wu
ua		wa
uo		wo
uai		wai
ui		wei
uan	→	wan
un		wen
uang		wang
ueng		weng

ü → y

ü	→	yu
ü		yu
üe		yue
üan	→	yuan
ün		yun

Practice: Spelling Rules

The syllables listed in the chart below are spelled wrong. Listen to the recordings for their pronunciations and follow the spelling rules discussed in the previous section to correct the misspellings.

	Incorrect	Correct		Incorrect	Correct		Incorret	Correct
1.	miou		**11.**	z		**21.**	uai	
2.	jiou		**12.**	ch		**22.**	ui	
3.	xiou		**13.**	sh		**23.**	uan	
4.	huei		**14.**	ie		**24.**	ü	
5.	cuen		**15.**	iou		**25.**	üe	
6.	suen		**16.**	iang		**26.**	üan	
7.	jü		**17.**	ing		**27.**	qiou	
8.	qüan		**18.**	iong		**28.**	kuen	
9.	xün		**19.**	ua		**29.**	guei	
10.	r		**20.**	uo		**30.**	zh	

Tones and Tone Marks

Four Tones

Chinese is a tonal language which has four pitched tones. Characters with the same sound but different tones express different meanings. For example:

1st tone	2nd tone	3rd tone	4th tone
qīng	qíng	qǐng	qìng
轻	晴	请	庆
(light)	(sunny)	(please)	(celebrate)

The diagram below illustrates the pitches and tone marks of the four Chinese tones. The first tone is a flat tone which is at a high pitch. The second tone is a rising tone which rises from a middle pitch to a high pitch. The third tone is a low rising tone which starts at a low pitch, then goes to the lowest pitch, and then rises to a high pitch. The fourth tone is a down tone which starts from the highest pitch and ends at the lowest pitch.

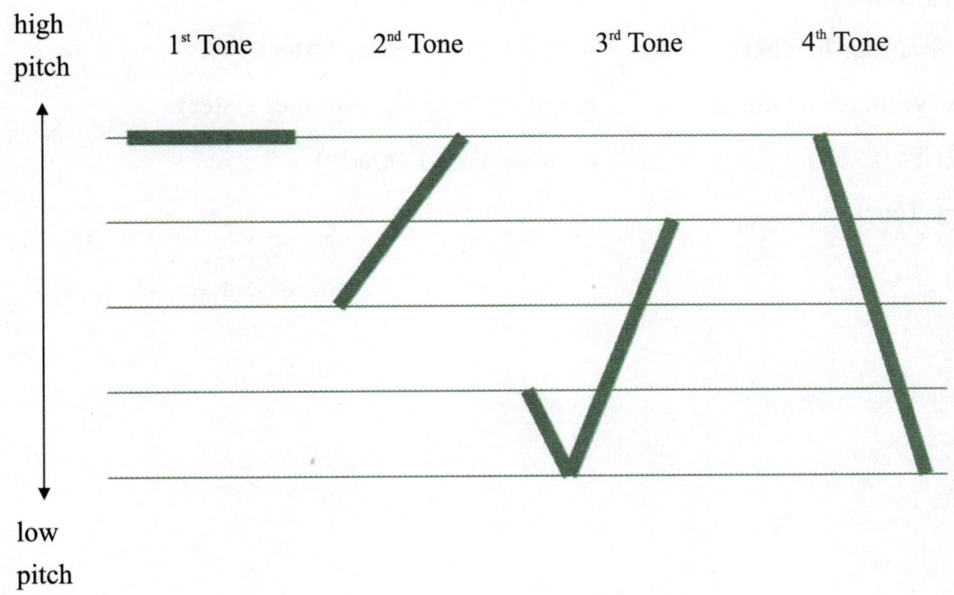

Practice: Four Tones

1 Repeat after the recordings of the four tones in each of the items listed below

	1st tone	2nd tone	3rd tone	4th tone
(1)	bā	bá	bǎ	bà
(2)	mā	má	mǎ	mà
(3)	fā	fá	fǎ	fà

2 Listen to the recordings and circle the tone pronounced in each item below

(1) dā dà (2) tú tǔ (3) né nè (4) lā lǎ

(5) gē gé (6) jī jǐ (7) qí qì

The Neutral Tone

The neutral tone is pronounced softly, but whose pitch is determined by the tones of the adjacent syllables. Neutral tones are usually unmarked. Listen to the recordings and try to imitate the sounds.

1. bàba 爸爸 (father)
2. māma 妈妈 (mother)
3. gēge 哥哥 (older brother)
4. jiějie 姐姐 (older sister)
5. dìdi 弟弟 (younger brother)
6. mèimei 妹妹 (younger sister)
7. hǎo ba 好吧 (O.K.)
8. hǎo le 好了(ready)
9. chēzi 车子 (car)

Tone Change Rules

Tone Change Rule 1

When there are two 3rd tones, the first 3rd tone is pronounced as a 2nd tone.

3rd + 3rd	pronounced as →	2nd + 3rd	
qǐng zǒu		qíng zǒu	请走 (please go)
nǐ hǎo		ní hǎo	你好 (hi)
lǎohǔ	→	láohǔ	老虎 (tiger)
lǎoshǔ		láoshǔ	老鼠 (mouse)

Tone Change Rule 2

If a 3rd tone is followed by a 1st, 2nd, or 4th tone, the 3rd tone is pronounced as a half 3rd tone. A half 3rd tone falls in pitch, but does not rise in pitch.

3rd + 1st, 2nd, or 4th	pronounced as →	half 3rd + 1st, 2nd, or 4th (the tone falls, but does not rise)	
dǎ qiú		dǎ qiú	打球 (batting)
zǒu lù		zǒu lù	走路 (walk)
pǎo bù		pǎo bù	跑步 (run)
xiě zì	→	xiě zì	写字 (write)

Tone Change Rule 3

When the character 不(bù) is followed by another 4th tone, 不 is pronounced as the 2nd tone.

bù + 4th	pronounced as →	bú + 4th	
bù shì		bú shì	不是 (not)
bù yào		bú yào	不要 (don't want)
bù qù	→	bú qù	不去 (don't go)
bù zuò		bú zuò	不做 (don't do)

Tone Change Rule 4

When the character 一 (yī) is followed by 1st, 2nd, or 3rd tones, 一 is pronounced as the 4th tone.

yī + 1st, 2nd, or 3rd	pronounced as →	yì + 1st, 2nd, or 3rd	
yī quān		yì quān	一圈 (a circle)
yī zhāng	→	yì zhāng	一张 (a piece)
yī nián		yì nián	一年 (a year)
yī diǎn		yì diǎn	一点 (a dot)

Tone Change Rule 5

When the character 一 (yī) is followed by a 4th tone, 一 is pronounced as the 2nd tone.

yī + 4th	pronounced as →	yí + 4th	
yīpiàn		yípiàn	一片 (a piece)
yībù		yíbù	一步 (a step)
yīxià		yíxià	一下 (a brief moment)
yīcì		yícì	一次 (once)

Practice: Tone Change Rules

The following syllables follow the tone change rules. Work in pairs. Read the syllables to each other. After you have completed your task, listen to and repeat after the recordings for more practice.

(1) nǐ hǎo (2) lǎoshǔ (3) xiě zì (4) zǒu lù (5) bù shì

(6) bù zuò (7) yī diǎn (8) yī quān (9) yīxià (10) yī cì

Tone Mark Rules

Tone Mark Rule 1

The tone mark is put above the vowel if there is only a single final.

bà	wǒ	nǐ
lè	qù	yī

Tone Mark Change Rule 2

The tone mark is put above the first vowel if there are two vowels. However, the tone mark is put above the second vowel if the first vowel is "i" or "u".

mèi	gōu	jiě
jiù	xióng	zhuǎn

Tone Mark Change Rule 3

The small dot above the "i" is omitted when a tone mark is written on "i".

Examples:

huī	chí	lín
yìn	qí	jī

Practice: Tone Mark Rules

Write the syllables with the tone marks on the right vowels. Item (1) is done for you as an example.

(1) wan, 2nd: wán　　(2) jue, 4th: ______　　(3) yin, 1st: ______

(4) shuo, 1st: ______　　(5) zhong, 3rd: ________　　(6) shi, 4th: ______

(7) xiong, 2nd: ______　　(8) gui, 3rd: ______　　(9) mei, 2nd: ______

Practice Chinese Pronunciations with Chinese Nursery Rhymes

Work in pairs. Take turns to read the following Chinese nursery rhymes.

1 小皮球 (Small Ball)

Xiǎo píqiú, xiāngjiāo yóu, 小皮球，香蕉油，
(Small ball, banana oil)
Mǎn dì kāi huā èr shí yī. 满地开花二十一。
(Flowers bloom on the ground, 21)
Èr wǔ liù, èr wǔ qī, 二五六，二五七，
(256, 257)
Èr bā, èr jiǔ, sān shí yī. 二八、二九、三十一。
(28, 29, 31)

2 三轮车 (Tricycle Runs Fast)

Sānlúnchē, pǎo de kuài, 三轮车，跑得快，
(Tricycle runs fast)
Hòumiàn zuò ge lǎo tàitai. 后面坐个老太太。
(An elderly woman is in the back seat)
Yào wǔ máo, gěi yī kuài, 要五毛，给一块，
(Asked for 50 cents, she gives 1 dollar)
Nǐ shuō qíguài bù qíguài? 你说奇怪不奇怪？
(Don't you agree it is strange?)

第二课 Lesson 2

汉字系统 The Chinese Writing System

Strokes

The Chinese writing system consists of three tiers: character, radical, and stroke. Strokes are the most basic units from which radicals and characters are created. There are 24 basic strokes, among which eight basic strokes are most commonly used.

Practice

The eight commonly used basic strokes are presented in the first boxes below. Practice writing the basic strokes in the blank boxes following the models of the strokes.

1 横 héng horizontal stroke

一	一	一							

2 竖 shù vertical stroke

丨	丨	丨							

3 撇 piě left leg

丿	丿	丿							

4 捺 nà right leg

㇏	㇏	㇏							

5 提 tí rising stroke

㇀	㇀	㇀							

6 点 diǎn dot stroke

丶	丶	丶							

7 钩 gōu hook stroke

亅	亅	亅							

8 折 zhé bend stroke

㇕	㇕	㇕							

The character, 永, consists of the eight most commonly used basic strokes. Practice how to write the eight basic strokes to make the character 永 in the boxes below.

永 

Stroke Order Rules

One should follow certain stroke order rules to write Chinese characters. Once one makes a habit of writing Chinese characters following the stroke order rules, one will find that writing Chinese characters is not a difficult task.

Rule 1: Top to bottom

早 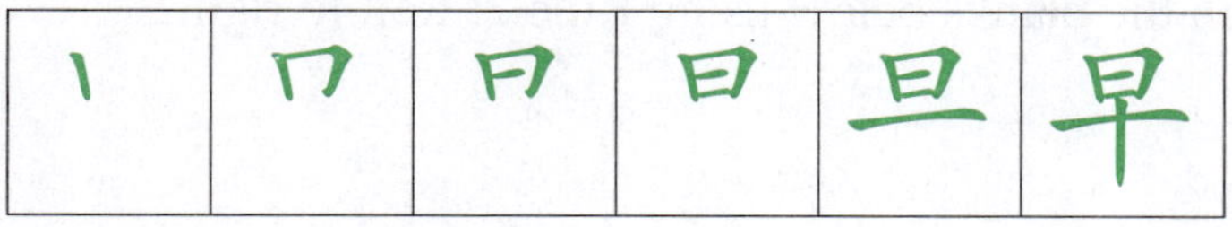

Rule 2: Left to right

什 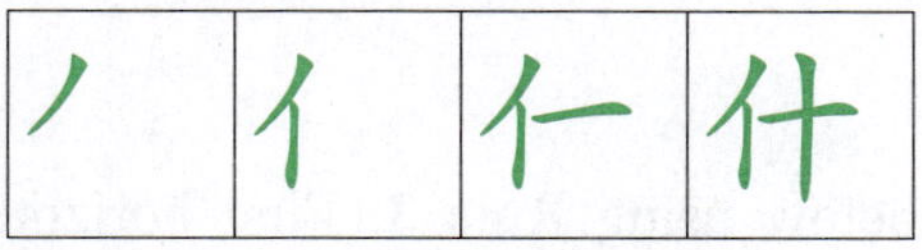

Rule 3: First horizontal, then vertical

十 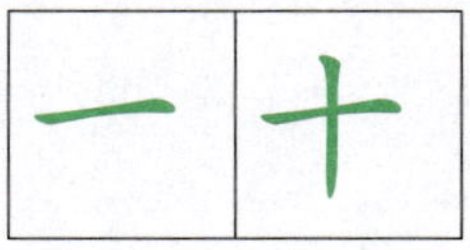

Rule 4: Outside first, then inside, then close the bottom side last

目 

Rule 5: Left leg first, then right leg

八 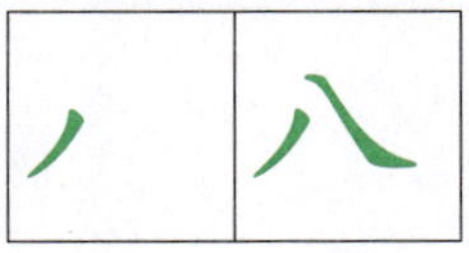

Rule 6: Middle first, then the sides

水 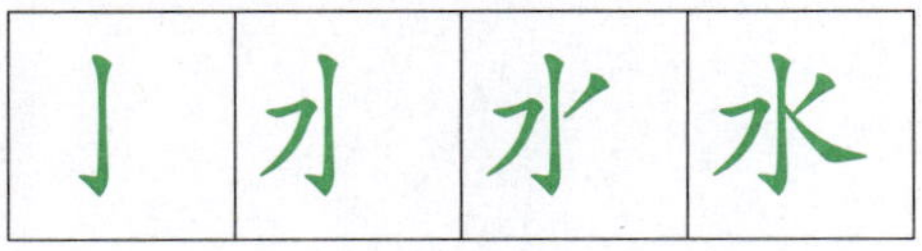

Practice

Write the following character 4 times in the blanks below using Rule 1: Top to bottom

古

古				

Write the following character 4 times in the blanks below using Rule 2: Left to right

叶

叶				

Write the following character 4 times in the blanks below using Rule 3: First horizontal, then vertical

十

十				

Write the following character 4 times in the blanks below using Rule 4: Outside first, then inside, then close the bottom side last

日

日				

Write the following character 4 times in the blanks below using Rule 5: Left leg first, then right leg

六

六				

Write the following character 4 times in the blanks below using Rule 6: Middle first, then the sides

小

小				

Character Structures

Radicals are the basic components in characters. There are more than 200 radicals in Chinese dictionaries. Some radicals can be used independently as characters. Such characters are called **simple characters**. A **compound character** is a combination of a radical with other components.

Listed below are examples of (1) simple characters: radicals used independently as characters, and (2) compound characters: radicals combined with other components

1 Simple Characters

人 [rén] (people)	女 [nǚ] (female)	心 [xīn] (heart)	门 [mén] (door)	车 [chē](car)
口 [kǒu] (mouth)	木 [mù] (wood)	雨 [yǔ] (rain)	马 [mǎ] (horse)	火 [huǒ] (fire)

2 Compound Characters

(radical: 人) 你 [nǐ] (you)	(radical: 女) 妈 [mā] (mother)	(radical: 心) 您 [nín] (you)	(radical: 门) 问 [wèn] (ask)
(radical: 车) 转 [zhuǎn] (turn)	(radical: 口) 叫 [jiào] (call)	(radical: 木) 树 [shù] (tree)	(radical: 雨) 雪 [xuě] (snow)

(radical: 马)
骂
[mà] (scold)

(radical: 火)
炒
[chǎo] (fry)

Practice

The characters listed below are compound characters. Each of the compound characters below includes one of the ten radicals introduced above. Identify the radical in each compound character by circling the radical and writing it down in the blank provided above the character.

(radical: ____) 吃 [chī] (eat)	(radical: ____) 雷 [léi] (thunder)	(radical: ____) 枝 [zhī] (branch)	(radical: ____) 灶 [zào] (stove)
(radical: ____) 间 [jiān] (room)	(radical: ____) 什 [shén] (what)	(radical: ____) 妹 [mèi] (younger sister)	(radical: ____) 怨 [yuàn] (complaint)
(radical: ____) 轨 [guǐ] (track)	(radical: ____) 驰 [chí] (speed up)	(radical: ____) 可 [kě] (can)	(radical: ____) 需 [xū] (need)
(radical: ____) 杖 [zhàng] (stick)	(radical: ____) 灸 [jiǔ] (cauterize)	(radical: ____) 闪 [shǎn] (flash)	(radical: ____) 亿 [yì] (billion)
(radical: ____) 姐 [jiě] (older sister)	(radical: ____) 忘 [wàng] (forget)	(radical: ____) 轧 [yà] (roll)	(radical: ____) 驱 [qū] (drive)

Compound Character Structures

Listed below are some common variations of compound character structures which have one radical and one or more other components.

Two part Structures

1. Left & Right:

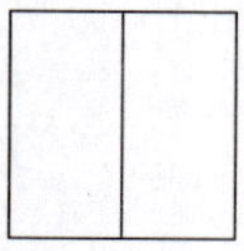

Example: 你

2. Top & Bottom:

Example: 古

3. Top, Left, & Bottom:

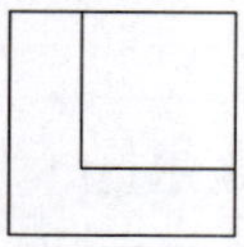

Example: 区

4. Box

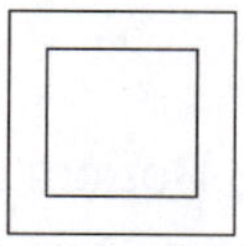

Example: 图

Three part Structures

1. Left, Middle, & Right:

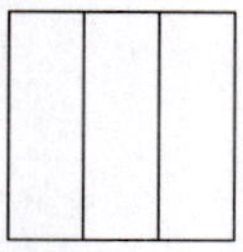

Example: 街

2. Top, Middle, & Bottom:

Example: 鼻

3. Left, Top, & Bottom:

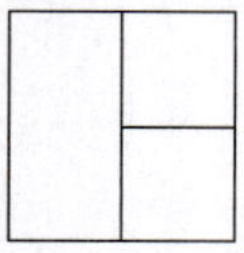

Example: 唱

4. Top, Left, & Right:

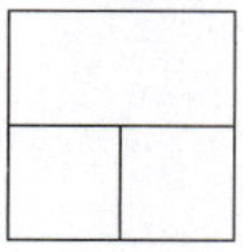

Example: 花

Practice

Analyze the structure of each of the compound characters listed below and draw a line between the character and its correct structure in the chart.

(1) Two part Structures

Left & Right	Top & Bottom	Top, Left, & Bottom	Box
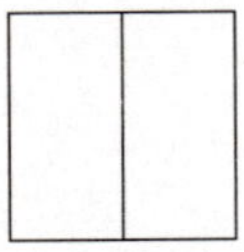	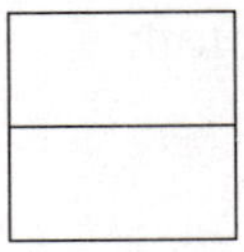	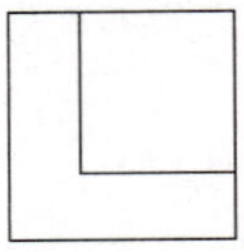	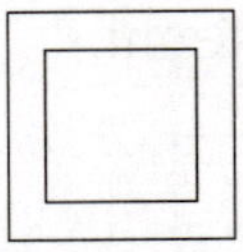

这　　爸　　妈　　国

(2) Three part Structures

Left, Middle, & Right	Top, Middle, & Bottom	Left, Top, & Bottom	Top, Left, & Right
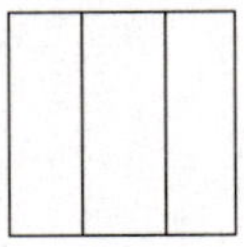	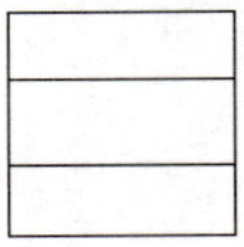	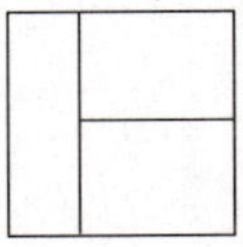	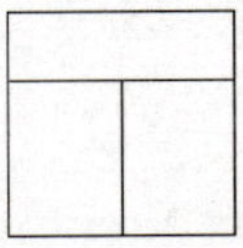

黄　　脚　　筷　　鞋

第三课 Lesson 3

很高兴认识你！
It's a Pleasure to Make Your Acquaintance!

沟通任务 Communication Tasks

- *Greeting your teacher and classmates*
- *Indicating an individual is a new student or is a new teacher*
- *Asking for someone's name*
- *Asking where a person is from*
- *Expressing your pleasure to meet new people*

课前讨论 Warm-up Discussion Questions

1. What is your name? Is your name a commonly used name?
2. Which country are you from? Which state or city are you from?

生词 Vocabulary

1 老师	老師	lǎoshī	teacher	*n.*

◎ 老师好！(Hello, teacher!)

2 学生	學生	xuéshēng	student	*n.*
们	們	men	plural for pronouns and human nouns	*suff.*
		◎ 学生们好！(Hello, students!)		
3 你	你	nǐ	you	*pron.*
		你好！(Hi!)		
4 我	我	wǒ	I, me	*pron.*
		◎ 我是小明。(I am Xiaoming.)		
5 您	您	nín	you (polite form)	*pron.*
		◎ 您好！(Hello!)		
6 好	好	hǎo	good	*adj.*
		◎ 你好！(Hi!)		
7 大家	大家	dàjiā	everyone	*pron.*
		◎ 大家好！(Hello, everyone!)		
8 叫	叫	jiào	to call	*v.*
		◎ 我叫爱华。(My name is Aihua.)		
9 什么	什麼	shénme	what	*pron.*
		◎ 你叫什么名字？(What is your name?)		
10 名字	名字	míngzi	name	*n.*
		◎ 你叫什么名字？(What is your name?)		
11 新	新	xīn	new	*adj.*
		◎ 我是新学生。(I am a new student.)		
12 请问	請問	qǐngwèn	May I ask... (to show politeness when asking a question)	*v.*
		◎ 请问，你叫什么名字？(May I ask, what is your name?)		
13 是	是	shì	to be	*v.*
		◎ 我是老师。(I am a teacher.)		

14 哪	哪	nǎ	which	*pron.*
		◎ 你是哪国人?(Which country are you from?)		
15 人	人	rén	people, person	*n.*
		◎ 我是中国人。(I am Chinese.)		
16 哪里	哪裏	nǎlǐ	where	*pron.*
		◎ 你是美国哪里人?(Where in the United States are you from?)		
17 他	他	tā	he, him	*pron.*
		◎ 他是美国人。(He is American.)		
18 她	她	tā	she, her	*pron.*
		◎ 她是中国人。(She is Chinese.)		
19 很	很	hěn	very	*adv.*
		◎ 我很好。(I am fine.)		
20 高兴	高興	gāoxìng	glad, happy	*adj.*
		◎ 我很高兴。(I am very happy.)		
21 认识	認識	rènshi	to know, to be acquainted	*v.*
		◎ 很高兴认识你!(It's a pleasure to make your acquaintance.)		
22 也	也	yě	also, too	*adv.*
		◎ 他也是中国人。(He is also Chinese.)		

专有名词 Proper Nouns

1 美国	美國	Měiguó	The United States of America	*n.*
		◎ 我是美国人。(I am American.)		
2 德州	德州	Dézhōu	Texas state	*n.*
		◎ 我是德州人。(I am Texan.)		
3 加州	加州	Jiāzhōu	California state	*n.*
		◎ 我是加州人。(I am Californian.)		

补充词语 Supplementary Vocabulary

1 中国	中國	Zhōngguó	China	*n.*
2 加拿大	加拿大	Jiānádà	Canada	*n.*
3 犹他州	猶他州	Yóutāzhōu	Utah	*n.*
4 北京	北京	Běijīng	Beijing	*n.*
5 上海	上海	Shànghǎi	Shanghai	*n.*
6 多伦多	多倫多	Duōlúnduō	Toronto	*n.*

沟通任务 Communication Tasks

I Greeting your teacher and classmates

Greet your teacher:

______老师，您好！(Hello, __________!)

Greet a classmate:

__________，你好！(Hello, __________!)

Greet the class:

大家好！(Hello everyone!)

Attention:

1. 您好 vs. 你好：

   你好 is the most frequently used greeting word.

   您好 is a polite form used when greeting an elderly person or someone who has a higher social status.

2. Chinese names: A Chinese name is written with the surname first and the given name next (last name first name). For example, a person whose surname is 王(Wáng) and given name is 小强(Xiǎoqiáng) is named 王小强(Wáng Xiǎoqiáng) in Chinese.

Examples:

王老师，您好！(Hello, teacher Wang!)

李小明，你好！(Hello, Xiaoming Li!)

大家好！(Hello everyone!)

Practice: Greet your teacher and classmates in Chinese.

II Indicating an individual is a new student or is a new teacher

__________ 是新老师。(__________ am/is a new teacher.)

__________ 是新学生。(__________ am/is a new student.)

Examples:

王老师是新老师。(Teacher Wang is a new teacher.)

我是新学生。(I am a new student.)

他是新学生。(He is a new student.)

Practice: Are there new students in your class? Identify them by using the structure taught above.

III Asking for someone's name

Question: 请问你叫什么名字？(What is your name?)

Answer: 我叫__________。(My name is __________.)

Attention:

1. 请问 is usually used before asking a question to show politeness.

Example:

Question: 请问你叫什么名字？(What is your name?)

Answer: 我叫李小明。(My name is Xiaoming Li.)

Practice: Ask your classmates what their Chinese names are.

IV Asking where a person is from

Question 1: 你是哪国人？(Which country are you from?)

Answer 1: 我是__________人。(I am from __________.)

Question 2: 你是__________哪里人？(Where in __________ are you from?)

Answer 2: 我是__________国__________人。

(I am from __________, __________.)

Examples:

Question 1: 你是哪国人？(Which country are you from?)

Answer 1: 我是美国人。(I am from the United States.)

Question 2: 你是美国哪里人？(Where in the United States are you from?)

Answer 2: 我是美国加州人。(I am from California, in the United States.)

Practice: Ask your classmates where they are from.

V Expressing your pleasure to meet new people

A: 很高兴认识你。(It's a pleasure to make your acquaintance.)

B: 我也很高兴认识你。(The pleasure is mine.)

Practice: Tell your new classmates, "It's a pleasure to make your acquaintance."

情景(一) Scenario 1

在中文班…… In a Chinese class...

李小明： 王老师，您好！我叫李小明。我是新学生。

Lǐ Xiǎomíng：Wáng Lǎoshī, nín hǎo! Wǒ jiào Lǐ Xiǎomíng. Wǒ shì xīn xuéshēng.

王老师： 李小明，你好！

Wáng lǎoshī: Lǐ Xiǎomíng, nǐ hǎo!

学生们： 李小明，你好！

Xuéshēngmen: Lǐ Xiǎomíng, nǐ hǎo!

李小明： 大家好！

Lǐ Xiǎomíng: Dàjiā hǎo!

Comprehension Questions

新学生叫什么名字?

情景(二) Scenario 2

中文课下课后…… Right after the Chinese class...

李小明: 你好! 我叫李小明。请问你叫什么名字?

Lǐ Xiǎomíng: Nǐ hǎo! Wǒ jiào Lǐ Xiǎomíng. Qǐngwèn nǐ jiào shénme míngzi?

张爱华: 你好,李小明。我叫张爱华。

Zhāng Àihuá: Nǐ hǎo, Lǐ Xiǎomíng. Wǒ jiào Zhāng Àihuá.

李小明: 你是哪国人?

Lǐ Xiǎomíng: Nǐ shì nǎ guó rén?

张爱华: 我是美国人。

Zhāng Àihuá: Wǒ shì Měiguó rén.

李小明: 你是美国哪里人?

Lǐ Xiǎomíng: Nǐ shì Měiguó nǎlǐ rén?

张爱华: 我是美国加州人。

Zhāng Àihuá: Wǒ shì Měiguó Jiāzhōu rén.

李小明: 我是美国德州人。

Lǐ Xiǎomíng: Wǒ shì Měiguó Dézhōu rén.

张爱华: 很高兴认识你。

Zhāng Àihuá: Hěn gāoxìng rènshi nǐ.

李小明: 我也很高兴认识你。

Lǐ Xiǎomíng: Wǒ yě hěn gāoxìng rènshi nǐ.

Comprehension Questions

1. 张爱华是哪国人?
2. 李小明是美国哪里人?

口语沟通活动 Oral Communication Activities

I. 情况：中文班有五个新学生。他们的名字是王小强 (Wáng Xiǎoqiáng)、林文 (Lín Wén)、陈龙 (Chén Lóng)、李丽 (Lǐ Lì)和王文丽 (Wáng Wénlì).

Situation: There are five new students in Chinese class. Their names are 王小强 (Wáng Xiǎoqiáng), 林文 (Lín Wén), 陈龙 (Chén Lóng), 李丽(Lǐ Lì), and 王文丽 (Wáng Wénlì).

任务：小组练习。每个组员扮演新学生之一。用下面列出的会话结构做角色扮演。彼此打招呼并问名字。

Task: Work in small groups. Each group member takes a role of the five new students. Use the conversation structure listed below to greet each other and ask for each others' names.

会话结构：

Conversation Structure:

林　文：你好。请问你叫什么名字？

王小强：你好。我叫<u>王小强</u>。请问你叫什么名字？

林　文：我叫<u>林文</u>。

王小强：很高兴认识你。

林　文：我也很高兴认识你。

II. 情况：这五个新学生是不同国家和城市的人。这些资料列在下表：

Situation: The five new students are from different states/cities and different countries. Their information is listed below:

姓名 Name	国家 Country	州/城市 State or City
王小强	美国	德州
林文	美国	犹他州
陈龙	中国	北京
李丽	中国	上海
王文丽	加拿大	多伦多

任务：小组练习。用下面列出的会话结构做角色扮演。问对方是哪里人。

Task: Work in the same small groups. Use the conversation structure listed below to find out where each new student is from.

会话结构：

Conversation Structure:

林　文：你是哪国人？

王小强：我是美国人。

林　文：你是美国哪里人？

王小强：我是美国德州人。

读写沟通活动 Literacy Communication Activities

I. 问你的同学和老师是哪里人。把答案写在下面的表格中。如果你不知道怎么用中文写，可以写拼音。

Find out where your teacher and each of your classmates are from. Write down their names and all the answers you get. Write in pinyin if you do not know how to write the characters. Report your findings to your class.

姓名 Name	国家 Country	州/城市 State or City

II. 用"我是country + state or city 人"这个句型在纸条上写出你是哪里人。把纸条折好交给老师。你的老师会请学生念出每个纸条上写的句子。班上学生得猜每个纸条是谁写的。

Write a note indicating where you are from using the structure "我是country + state or city 人". Fold the note and give it to your teacher. Your teacher will call students to read the notes each of the students has turned in. The class will guess who wrote each of the notes.

例子： Examples of the note:

1.

我是美国加州人。

2.

我是中国上海人。

讨论Discussion

Chinese Surnames

According to the Ministry of Public Security of the People's Republic of China, there are currently around 4700 Chinese surnames being used. The most common surnames in use in China are 王(Wang), followed by 李(Li) and 张(Zhang). Approximately 92.9 million people in China use the surname 王, 92 million people in China use the surname 李, and 87.5 million people in China use the surname 张.

Listed below are the top 10 most common surnames in China. More than 20 million people use each of the 10 surnames.

王 Wáng	李 Lǐ	张 Zhāng	刘 Liú	陈 Chén
杨 Yáng	黄 Huáng	赵 Zhào	吴 Wú	周 Zhōu

Discussion Question:

1. Did your teacher give you a Chinese name? What is your Chinese surname? Is your Chinese surname one of the 10 most commonly used surnames?
2. How many students in class have the top 10 common surnames? Try to identify them in class.

第四课
Lesson 4

我爱我的家
I Love My Family

沟通任务 Communication Tasks

- *Asking how many people are in someone's family*
- *Using 你呢？(How about you?) as a follow up question*
- *Asking about someone's family*
- *Asking if someone has siblings*
- *Telling people how much you love your family*

课前讨论 Warm-up Discussion Questions

1. How many people are in your family? Who are in your family?
2. Do you have any siblings?
3. Do you have any pets?

生词 Vocabulary

1 家	家	jiā	home, family	*n.*
		◎ 我家人是中国人。(My family members are Chinese.)		
2 有	有	yǒu	to have	*v.*
		◎ 我有中国老师。(I have teachers from China.)		
3 没有	没有	méiyǒu	to not have	*v.*
		◎ 我没有中国老师。(I don't have teachers from China.)		
4 几	幾	jǐ	how many	*num.*
		◎ 你有几个学生? (How many students do you have?)		
5 个	個	gè	general classifier	*m.w.*
		◎ 我家有三个人。(There are three people in my family.)		
6 呢	呢	ne	question particle	*part.*
		◎ 我是新学生。你呢? (I am a new student. How about you?)		
7 谁	誰	shéi	who	*pron.*
		◎ 你是谁? (Who are you?)		
8 爸爸	爸爸	bàba	father	*n.*
		◎ 我爸爸是老师。(My father is a teacher.)		
9 妈妈	媽媽	māma	mother	*n.*
		◎ 我妈妈很好。(My mother is fine.)		
10 哥哥	哥哥	gēge	older brother	*n.*
		◎ 我哥哥是学生。(My elder brother is a student.)		
11 姐姐	姐姐	jiějie	older sister	*n.*
		◎ 我有一个姐姐。(I have an elder sister.)		
12 弟弟	弟弟	dìdi	younger brother	*n.*
		◎ 我有一个弟弟。(I have a younger brother.)		

13 妹妹	妹妹	mèimei	younger sister	*n.*
		◎ 我没有妹妹。(I don't have a younger sister.)		
14 和	和	hé	and	*conj.*
		◎ 我爸爸和妈妈是老师。(My father and mother are teachers.)		
15 兄弟姐妹	兄弟姐妹	xiōngdì jiěmèi	siblings	
		◎ 我有三个兄弟姐妹。(I have three siblings.)		
16 吗	嗎	ma	particle used to form a yes or no question	*part.*
		◎ 你有三个哥哥吗?(Do you have three brothers?)		
17 但是	但是	dànshì	but	*conj.*
		◎ 我没有妹妹,但是我有一个姐姐。(I don't have a younger sister, but I have an elder sister.)		
18 狗	狗	gǒu	dog	*n.*
		◎ 我没有狗。(I don't have a dog.)		
19 只	隻	zhī	classifier for animals	*m.w.*
		◎ 我有一只狗。(I have a dog.)		
20 的	的	de	possessive particle	*part.*
		◎ 我的爸爸是老师。(My father is a teacher.)		
21 爱	愛	ài	to love	*v.*
		◎ 我爱我的狗。(I love my dog.)		

补充词语 Supplementary Vocabulary

1 一	一	yī	one	*num.*
2 二	二	èr	two	*num.*
3 三	三	sān	three	*num.*

4 四	四	sì	four	*num.*
5 五	五	wǔ	five	*num.*
6 六	六	liù	six	*num.*
7 七	七	qī	seven	*num.*
8 八	八	bā	eight	*num.*
9 九	九	jiǔ	nine	*num.*
10 十	十	shí	ten	*num.*
11 猫	貓	māo	cat	*n.*

沟通任务 Communication Tasks

I Asking how many people are in someone's family

Question: 你家有几个人？(How many people are in your family?)

Answer: 我家有____个人。(There are ____ people in my family.)

Example:

Question: 你家有几个人？(How many people are in your family?)

Answer: 我家有三个人。(There are three people in my family.)

Attention:

1. See Supplementary Vocabulary for Chinese numbers 1—10.
2. 二 [èr] vs. 两 [liǎng]: Both 二 and 两 mean two. When the number two is followed by a classifier, such as 个, use 两 instead of 二. For example, 两个人。

Practice: Ask a classmate sitting next to you how many people are in his or her family.

II Using 你呢? (How about you?) as a follow up question

A: __________?

B: __________, 你呢？(How about you?)

A: __________。

Examples:

A: 你家有几个人？(How many people are in your family?)

B: 我家有三个人。你呢？(There are three people in my family. How about you?)

A: 我家有五个人。(There are five people in my family.)

Practice: Find a partner and ask a question in Chinese. Your partner needs to answer your question and ask "你呢？" in order for you to answer the same question.

III Asking about someone's family

Question: 你家有谁？(Who is in your family?)

Answer: 我家有______、______和______。

(My family is made up of ______, ______ and ______.)

Example:

Question: 你家有谁？(Who is in your family?)

Answer: 我家有爸爸、妈妈和我。(My family is made up of my father, my mother, and me.)

Practice: Ask a classmate who is in his or her family.

IV Asking if someone has siblings

Question: 你有兄弟姐妹吗？(Do you have siblings?)

Positive Answer: 有，我有____个哥哥、____个姐姐、____个弟弟和____个妹妹。

(Yes, I have ____ elder brothers, ____ elder sisters, ____ younger brothers, and ____ younger sisters.)

Or

Negative Answer: 没有，我没有兄弟姐妹，但是我有____只狗。

(No, I do not have any siblings, but I do have ____ dog(s).)

Examples:

A: 你有兄弟姐妹吗？(Do you have siblings?)

B: 有，我有三个哥哥、两个姐姐、三个弟弟和两个妹妹。你呢？

(Yes, I have three older brothers, two older sisters, three younger brothers, and two younger sisters. How about you?)

A: 没有，我没有兄弟姐妹，但是我有两只狗。

(No, I do not have any siblings, but I do have two dogs.)

Practice: Ask a classmate if he or she has siblings.

V Telling people how much you love your family

我家有______________。我爱我的家！

(My family is made up of ______________. I love my family!)

Attention:

1. 我家 vs. 我的家：Both 我家 and 我的家 mean "my family". The possessive particle 的 is optional when the speaker refers to a person. However, 的 must be used in other cases. For example, 我的狗 (My dog).

Example:

我家有爸爸、妈妈、我和我的狗。我爱我的家！

(My family is made up of my father, my mother, me, and my dog. I love my family!)

Practice: Tell the class who is in your family and how much you love your family.

情景(一) Scenario 1

李小明：爱华，你家有几个人？

Lǐ Xiǎomíng: Àihuá, nǐ jiā yǒu jǐ ge rén?

张爱华：我家有三个人。你呢？

Zhāng Àihuá: Wǒ jiā yǒu sān ge rén. Nǐ ne?

李小明：我家有六个人。

Lǐ Xiǎomíng: Wǒ jiā yǒu liù ge rén.

张爱华：你家有谁？

Zhāng Àihuá: Nǐ jiā yǒu shéi?

李小明：我家有爸爸、妈妈、两个姐姐、一个弟弟和我。你呢？

Lǐ Xiǎomíng: Wǒ jiā yǒu bàba, māma, liǎng ge jiějie, yí ge dìdi hé wǒ. Nǐ ne?

张爱华： 我家有爸爸、妈妈和我。

Zhāng Àihuá: Wǒ jiā yǒu bàba, māma hé wǒ.

Comprehension Questions

1. 张爱华家有几个人？
2. 李小明家有谁？

情景（二）Scenario 2

王文丽： 家玲，你有兄弟姐妹吗？

Wáng Wénlì: Jiālíng, nǐ yǒu xiōngdì jiěmèi ma?

吴家玲： 有，我有一个哥哥、一个姐姐、一个弟弟和一个妹妹。你呢？

Wú Jiālíng: Yǒu, wǒ yǒu yí ge gēge, yí ge jiějie, yí ge dìdi hé yí ge mèimei. Nǐ ne?

王文丽： 没有，我没有兄弟姐妹，但是我有一只狗。

Wáng Wénlì: Méiyǒu, wǒ méiyǒu xiōngdì jiěmèi, dànshì wǒ yǒu yì zhī gǒu.

我家有爸爸、妈妈、我和我的狗。我爱我的家！

Wǒ jiā yǒu bàba, māma, wǒ hé wǒ de gǒu. Wǒ ài wǒ de jiā!

Comprehension Questions

1. 吴家玲有兄弟姐妹吗？
2. 王文丽的家有谁？

口语沟通活动 Oral Communication Activities

I. 情况：下面有五张图。每个图里有一个家庭。

Situation: There are five pictures below. Each picture shows a different family.

任务：两人一组。用下面列出的会话结构轮流问对方每个家庭有几个人以及每个家庭有谁。

Task: Find a partner in class. Use the conversation structure listed below to take turns to ask each other how many family members are in each picture and who the family members are.

会话结构：

Conversation Structure:

你：李小明家有几个人？

你的伙伴：李小明家有三个人。

你：李小明家有谁？

你的伙伴：李小明家有爸爸、妈妈和李小明。

图一：刘中的家

图二：陈加加的家

图三：王小强的家

图四：李小明的家

图五：安安的家

II. 情况：你的老师和同学想多了解你和你的家人。你家有几个人？你家有谁？
Situation: Your teacher and classmates want to know more about you and your family. How many people are in your family? Who are your family members?

任务：画你的家人或带你家人的照片到班上。用下面的会话结构来介绍你家人。别忘了告诉大家你爱你的家人。
Task: Draw or bring a picture of your family. Use the following conversation structure to introduce your family to the class. Do not forget to tell the class how much you love your family!

会话结构：
Conversation Structure:
你：我家有______个人。我家有______、______、______、……我爱我的家！

读写沟通活动 Literacy Communication Activities

I. 在教室走动并一个一个问你的同学："你家有谁？"请把他们的答案记录在下面的表格中。例子请看表格第一行。
Walk around the classroom and ask your classmates one by one "你家有谁？". Record their names and their answers in the table listed below. See the example in the first row.

名字	爸爸	妈妈	哥哥	姐姐	弟弟	妹妹	狗	猫
陈小明	√	√	一个	0	两个	一个	0	一只
1.								
2.								
3.								
4.								
5.								
6.								
7.								
8.								
9.								

II. 用一小段话描述上面表格里某个同学家人的情况。这个同学的家有谁？
Write one of the classmates' family information recorded above in a short paragraph. Who is in the classmate's family?

例子: 陈小明
陈小明的家有爸爸、妈妈、两个哥哥、一个弟弟、一个妹妹、一只猫和陈小明。

在下面写出你同学家的情况：(Write your classmate's family information below:)
同学姓名：(Classmate's Name:)____________

____

____

讨论 Discussion

One-Child Policy in China

The one-child policy is called 计划生育政策 (jìhuà shēngyù zhèngcè) in Chinese. This policy was established in 1979 to limit China's population growth. Under this policy many couples may have only one child. However, it only applies to married urban couples, and allows exemptions for rural couples, ethnic minorities, and couples without any siblings themselves. The one-child policy is a way to prevent overpopulation in China.

Discussion Questions:

1. Do you agree with the one-child policy?
2. What could be some pros and cons of the results of the one-child policy?
3. If you were a Chinese government officer, what policies would you implement in order to solve the issue of overpopulation in China?

第五课
Lesson 5

我喜欢学中文
I Like Studying Chinese

沟通任务 Communication Tasks

- *Asking if someone likes to do certain things*
- *Describing what you do every day*
- *Making a suggestion and asking for an opinion*
- *Asking "why" questions*

课前讨论 Warm-up Discussion Questions

1. What languages do you like to study?
2. Do you like studying Chinese? What are some reasons you like studying Chinese?

生词 Vocabulary

1	喜欢	喜歡	xǐhuan	to like	*v.*

◎ 我喜欢中国人。(I like Chinese people.)

2 学	學	xué	to study	*v.*
		◎ 我学中文。(I study Chinese language.)		
3 中文	中文	Zhōngwén	Chinese language	*n.*
		◎ 我是中文老师。(I am a Chinese language teacher.)		
4 啊	啊	a	particle placed at the end of a sentence for emphasis	*part.*
		◎ 是啊！我有三个兄弟姐妹。(You are right! I have three siblings.)		
5 每	每	měi	every	*pron.*
		◎ 我每天学中文。(I study Chinese every day.)		
6 写	寫	xiě	to write	*v.*
		◎ 我每天写中文。(I write Chinese every day.)		
7 汉字	漢字	Hànzì	Chinese character	*n.*
		◎ 我每天写汉字。(I write Chinese characters every day.)		
8 读	讀	dú	to read	*v.*
		◎ 我每天读书。(I read every day.)		
9 书	書	shū	book	*n.*
		◎ 我喜欢读中文书。(I like to read Chinese books.)		
10 说	説	shuō	to speak	*v.*
		◎ 我每天说中文。(I speak Chinese every day.)		
11 我们	我們	wǒmen	we, us	*pron.*
		◎ 我们是兄弟姐妹。(We are siblings.)		
12 一起	一起	yìqǐ	together	*adv.*
		◎ 我们一起说中文。(We speak Chinese together.)		
13 练习	練習	liànxí	to practice	*v.*
		◎ 我们一起练习写汉字。(We practice writing Chinese characters together.)		

14 好吗	好嗎	hǎo ma	is it all right with someone (asking for agreement)	
		◎ 我们学中文好吗？ (Let's study Chinese, is that all right with you?)		
15 好	好	hǎo	great (expressing agreement)	adj.
		◎ 好！(Great!)		
16 为什么	為什麼	wèi shénme	why	
		◎ 为什么你喜欢学中文？ (Why do you like to study Chinese?)		
17 因为	因為	yīnwèi	because	conj.
		◎ 我学中文因为我妈妈说中文。 (I study Chinese because my mother speaks Chinese.)		
18 多	多	duō	many; a lot of	adj.
		◎ 我有很多妹妹。(I have many younger sisters.)		
19 朋友	朋友	péngyou	friend	n.
		◎ 我有很多中国朋友。(I have many Chinese friends.)		
20 都	都	dōu	all	adv.
		◎ 我们都说中文。(We all speak Chinese.)		

补充词语 Supplementary Vocabulary

1 英文	英文	Yīngwén	English language	n.
2 日文	日文	Rìwén	Japanese language	n.
3 韩文	韓文	Hánwén	Korean language	n.
4 阿拉伯文	阿拉伯文	Ālābówén	Arabic language	n.
5 西班牙文	西班牙文	Xībānyáwén	Spanish language	n.
6 法文	法文	Fǎwén	French language	n.
7 德文	德文	Déwén	German language	n.

8 去中国	去中國	qù Zhōngguó	to go to China	v.o.
9 打篮球	打籃球	dǎ lánqiú	to play basketball	v.o.
10 听音乐	聽音樂	tīng yīnyuè	to listen to music	v.o.

沟通任务 Communication Tasks

I Asking if someone likes to do certain things

Question: 你喜欢________吗？(Do you like ________?)

Positive Answer: 是啊！我很喜欢________。(Yes! I like ________ very much.)

Negative Answer: 我不喜欢 __________。(I do not like __________.)

Example:

Question: 你喜欢学中文吗？(Do you like studying Chinese?)

Positive Answer: 是啊！我很喜欢学中文。(Yes! I like studying Chinese very much.)

Negative Answer: 我不喜欢学中文。(I don't like studying Chinese.)

Practice: Ask a classmate if he or she likes studying Chinese.

II Describing what you do every day

我每天 ______________________。(I ______________________ every day.)

Examples:

我每天说中文。(I speak Chinese every day.)

我每天写汉字、读中文书。(I write Chinese characters and read Chinese books every day.)

Practice: Do you speak, read, or write Chinese every day? Tell your classmates if you do.

III Making a suggestion and asking for an opinion

Question: 我们一起__________好吗？

(Let's __________ together, is it all right with you?)

Positive Answer: 好啊！(Great!)

Negative Answer: 不好。(No.)

Example:

Question: 我们一起练习中文好吗？(Let's practice Chinese together, is it all right with you?)

Positive Answer: 好啊！(Great!)

Negative Answer: 不好。(No.)

Practice: Find a partner. Make the following suggestions to him or her in Chinese: 1. Let's speak Chinese together; 2. Let's write Chinese characters together; 3. Let's read Chinese books together. Are his or her answers positive or negative to the three suggestions?

IV Asking "why" questions

Question: 为什么你喜欢____________？(Why do you like ____________?)

Answer: 因为 ____________。(Because ____________.)

Example:

Question: 为什么你喜欢学中文？(Why do you like studying Chinese?)

Answer: 因为我有很多中国朋友。他们都说中文。

(Because I have many Chinese friends. They all speak Chinese.)

Practice: Ask your classmates why they like to study Chinese.

情景(一) Scenario 1

李小明: 你喜欢学中文吗？

Lǐ Xiǎomíng: Nǐ xǐhuan xué Zhōngwén ma?

张爱华: 是啊！我很喜欢学中文。我每天写汉字、读中文书。你呢？

Zhāng Àihuá: Shì a! Wǒ hěn xǐhuan xué Zhōngwén. Wǒ měi tiān xiě Hànzì, dú Zhōngwén shū. Nǐ ne?

李小明: 我也喜欢学中文。我每天说中文。

Lǐ Xiǎomíng: Wǒ yě xǐhuan xué Zhōngwén. Wǒ měi tiān shuō Zhōngwén.

张爱华: 我们一起练习中文好吗？

Zhāng Àihuá: Wǒmen yìqǐ liànxí Zhōngwén hǎo ma?

李小明: 好啊！

Lǐ Xiǎomíng: Hǎo a!

Comprehension Questions

1. 张爱华喜欢学中文吗?
2. 李小明喜欢学中文吗?

情景(二) Scenario 2

王文丽: 为什么你喜欢学中文?

Wáng Wénlì: Wèi shénme nǐ xǐhuan xué Zhōngwén?

吴家玲: 因为我有很多中国朋友。他们都说中文。你呢?

Wú Jiālíng: Yīnwèi wǒ yǒu hěn duō Zhōngguó péngyou. Tāmen dōu shuō Zhōngwén. Nǐ ne?

王文丽: 因为我爸爸妈妈是中国人。

Wáng Wénlì: Yīnwèi wǒ bàba māma shì Zhōngguó rén.

Comprehension Questions

1. 为什么王文丽喜欢学中文?
2. 为什么吴家玲喜欢学中文?

口语沟通活动 Oral Communication Activities

I. 情况:李小明在注册组工作。他的工作是帮助学生报名上学校的外语课。学校的外语课信息列在下面:

Situation: Li Xiaoming works for the registrar's office. His job is to help students sign up for the language courses offered in school. The school offers the following language courses:

外语课 Courses	
1. 英文	5. 阿拉伯文
2. 中文	6. 西班牙文
3. 日文	7. 法文
4. 韩文	8. 德文

任务:小组练习。一个人扮演李小明,其他人扮演要报名外语课的学生。李小明得用下面的会话结构问学生想学什么语言。

Task: Work in small groups. One of you acts as Li Xiaoming, and the rest act as students who need to sign up for language courses. Li Xiaoming needs to use the following conversation structure to find out what language courses students like to take.

会话结构:

Conversation Structure:

李小明:你喜欢学中文吗?

学生:是啊!我很喜欢学中文。

或

我不喜欢学中文。

如果学生不喜欢学某个语言,李小明得问他是否喜欢学其他语言。把学生的答案写在下表。

If a student does not like Chinese, Li Xiaoming needs to ask if that student likes studying other languages. Record the students' preferences in the chart below for the next activity.

学生姓名 Student Name	学生想学的语言 Courses the Student Likes to Study

李小明问了全部学生后，每个小组得在班上报告小组里的学生都想学什么语言。用以下句型：

After all students have told Li Xiaoming the languages they like to study, each small group needs to report to the class the language that each student likes to study. Please use the following structure:

________________ 喜欢学 ________________。

(学生姓名 student name)　　　　(语言 language)

II. 情况：下表内列出了几个你想跟你朋友一起做的有趣的活动

Situation: There are several fun activities listed below that you want to do with your friends together.

有趣的活动 Fun Activities	
活动 1 Activity 1	去中国 (go to China)
活动 2 Activity 2	打篮球 (play basketball)
活动 3 Activity 3	听音乐 (listen to music)

任务：用下列结构，“我们一起______好吗”问班上三个朋友他们想不想一起做。

Task: Use the structure, “我们一起______好吗” to ask three of your friends in class if they want to do the activities together.

会话结构：

Conversation Structure:

你：我们一起去中国好吗？

你朋友：好啊！

或

不好。

你和你朋友想一起做哪个活动？请在班上做口头报告。

Which activities do your friends and you like to do together? Please report the answers to the class.

读写沟通活动 Literacy Communication Activities

I. 李小明、张爱华和 王文丽都喜欢学中文。下面图片显示了他们每个人每天都怎么练习中文。看下面图片并在图片下写出他们都怎么练习中文。

Li Xiaoming, Zhang Aihua, and Wang Wenli all like studying Chinese. The pictures below show how each of these people practices Chinese every day. Look at the pictures and write down what each of the aforementioned people does to study Chinese.

图 1 李小明

李小明每天____________________。

图 2 张爱华

张爱华每天____________________。

图3 王文丽

王文丽每天____________________。

II. 你的一个朋友留了一张纸条给你，问你为什么学中文，以及你怎么练习中文。把你的回答写在下面纸条里。

A friend of yours left a note to you asking why you study Chinese and what you do to practice Chinese. Write your reply on the note below to your friend.

写下你的回复：(Write your reply:)

我每天____________________。
我喜欢学中文因为______________________________。

讨论Discussion

The Chinese Language

Chinese is considered one of the top 10 most influential languages in the world. It has more than a billion native and 20 million non-native speakers (Weber, 1997). There are many Chinese-speaking places in the world: (1) People's Republic of China (1.1 billion), (2) Taiwan (23 million), (3) Hong Kong (5 million), (4) Malaysia (4 million), and (5) Singapore (1 million).

The Modern Language Association (2009) conducted a foreign language enrollment survey of more than 2,500 U.S. universities and found that Chinese is one of the languages which has the highest growing enrollments. There were 61,000 U.S. college students studying Chinese in 2009. In addition, many American students are studying Chinese in China. In 2006, more than 11,000 students went to China (Open Doors, 2008).

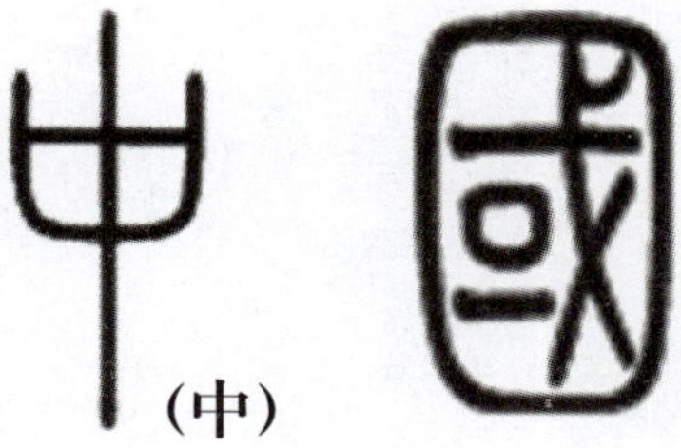

(中) (國)

Discussion Question:

Do you hope to travel and study in China one day? What cities in China would you like to visit? If you do not know any cities in China, ask your teacher to inform you about a few cities in China.

References:

Modern Language Association (2009) Enrollments in Languages Other Than English in United States Institutions of Higher Education, Fall 2009. Retrieved December 31, 2010, from http://www.mla.org/pdf/2009_enrollment_survey.pdf

Open Doors (2008) *Report on International Educational Exchange*. Sewickley, PA: Institute for International Education.

Weber, G. (1997) *Top Languages: The World's 10 Most Influential Languages Language Today*.

第六课 Lesson 6

明天你想做什么？
What Do You Want to Do Tomorrow?

沟通任务 Communication Tasks

- *Asking for people's activity plans*
- *Inviting others to do activities together*
- *Stating your opinion about your friend's activity proposal*
- *Making an activity proposal*
- *Accepting or declining an activity proposal*

课前讨论 Warm-up Discussion Questions

1. What kinds of activities do you do when you have free time?
2. Do you often invite your friends to do activities with you?

生词 Vocabulary

1	想	想	xiǎng	to want	*v.*

◎ 你想学中文吗？(Do you want to study Chinese?)

2 做	做	zuò	to do	*v.*
		◎ 你想做什么?(What do you want to do?)		
3 知道	知道	zhīdào	to know	*v.*
		◎ 我知道。(I know.)		
4 不	不	bù	not	*adv.*
		◎ 我不知道。(I don't know.)		
5 明天	明天	míngtiān	tomorrow	*n.*
		◎ 明天我想学中文。(I want to study Chinese tomorrow.)		
6 看电影	看電影	kàn diànyǐng	to see a movie	*v.o.*
		◎ 我想看电影。(I want to see a movie.)		
看	看	kàn	to see, to look	*v.*
电影	電影	diànyǐng	movie	*n.*
7 吃中国菜	吃中國菜	chī Zhōngguócài	to eat Chinese food	*v.o.*
		◎ 我想吃中国菜。(I want to eat Chinese food.)		
吃	吃	chī	to eat	*v.*
中国菜	中國菜	Zhōngguócài	Chinese food, Chinese dish	*n.*
8 打篮球	打籃球	dǎ lánqiú	to play basketball	*v.o.*
		◎ 你想打篮球吗?(Do you want to play basketball?)		
打	打	dǎ	to hit, to play	*v.*
篮球	籃球	lánqiú	basketball	*n.*
9 没关系	没關係	méi guānxi	never mind	
		◎ 没关系。(Never mind.)		
10 去	去	qù	to go	*v.*
		◎ 我们想一起去吃中国菜。(We want to go eat Chinese food together.)		

11 听音乐	聽音樂	tīng yīnyuè	to listen to music	v.o.
		◎ 你喜欢听什么音乐？(What music do you like to listen to?)		
听	聽	tīng	to listen	v.
音乐	音樂	yīnyuè	music	n.
12 怎么样	怎麼樣	zěnmeyàng	how	pron.
		◎ 我们听音乐怎么样？(How about we listen to music?)		
13 行	行	xíng	OK	v.
		◎ 行 (OK.)		
14 太好了	太好了	tài hǎo le	great	
		◎ 太好了。(Great.)		
15 见	見	jiàn	to see, to meet	v.
		◎ 明天见。(See you tomorrow.)		
16 游泳	游泳	yóu yǒng	to swim	
		◎ 我每天游泳。(I swim every day.)		
17 空儿	空儿	kòngr	free time	n.
		◎ 你有没有空儿？(Do you have free time?)		
18 今天	今天	jīntiān	today	n.
		◎ 我今天很好。(I am fine today.)		
19 后天	後天	hòutiān	day after tomorrow	n.
		◎ 后天他想读书。(He wants to read the day after tomorrow.)		
20 吧	吧	ba	particle placed at the end of a sentence to make a suggestion	part.
		◎ 我们学中文吧。(Let's study Chinese.)		
21 再见	再見	zàijiàn	goodbye	v.
		◎ 爸爸再见！(Goodbye father.)		

补充词语 Supplementary Vocabulary

1 泡茶	泡茶	pào chá	to make tea	v.o.
2 打麻将	打麻將	dǎ májiàng	to play mahjong	v.o.
3 逛夜市	逛夜市	guàng yèshì	to stroll through a night market	v.o.
4 做早操	做早操	zuò zǎocāo	to do morning exercises	v.o.
5 唱卡拉OK	唱卡拉OK	chàng kǎlā OK	to sing karaoke	v.o.
6 下棋	下棋	xià qí	to play chess	v.o.

沟通任务 Communication Tasks

I Asking for people's activity plans

Question: ________你想做什么? (What do you want to do ________?)
(date/time)

Answer: ________我想________、________和________。
(date/time)
(I want to ________, ________, and ________ ________.)
(date/time)

Attention:

1. Time words such as 今天, 明天 and 后天 can be placed before or after the subject.

For example:

"我今天学中文" and "今天我学中文" have the same meaning.

Example:

Question: 明天你想做什么? (What do you want to do tomorrow?)

Answer: 明天我想看电影、吃中国菜和打篮球。

(I want to see a movie, eat Chinese food and play basketball tomorrow.)

Practice: Find a partner in class and ask each other about your activity plans using the sentence patterns introduced above.

II Inviting others to do activities together

你想不想一起去？(Do you want to go together?)

Example:

明天我想看电影、吃中国菜和打篮球。你想不想一起去？

(I want to see a movie, eat Chinese food, and play basketball. Do you want to go together?)

Practice: Use the sentence patterns introduced above to tell your partner an activity plan and invite him or her to join you.

III Stating your opinion about your friend's activity proposal

我想________和________，但是我不想________。

(I want to ________ and ________, but I don't want to ________.)

Example:

我想看电影和吃中国菜，但是我不想打篮球。

(I want to see a movie and eat Chinese food, but I don't want to play basketball.)

Practice: Use the sentence structures introduced above to state your opinions about your partner's activity proposal. Tell him or her what activities you want or don't want to do.

IV Making an activity proposal

我们________怎么样？(How about we ________?)

Example:

我们听音乐怎么样？(How about we listen to music?)

Practice: Use the sentence structure above to suggest an activity for tomorrow to your partner.

V Accepting or declining an activity proposal

Accepting a suggestion:

行。(OK.)

好。(OK.)

太好了。(Great.)

Declining a suggestion:

我不想______。(I don't want to ______.)

我没有空。(I don't have free time.)

Example:

Question: 我们听音乐怎么样?(How about we listen to music?)

Positive Answer: 太好了!(Great!)

Negative Answer: 我不想听音乐。(I don't want to listen to music.)

Practice: Use the comments introduced above to show your agreement to do an activity together with your partner.

情景(一) Scenario 1

李小明: 爱华,明天你想做什么?

Lǐ Xiǎomíng: Àihuá, míngtiān nǐ xiǎng zuò shénme?

张爱华: 我不知道。你呢?

Zhāng Àihuá: Wǒ bù zhīdào. Nǐ ne?

李小明: 明天我想看电影、吃中国菜和打篮球。你想不想一起去?

Lǐ Xiǎomíng: Míngtiān wǒ xiǎng kàn diànyǐng, chī Zhōngguó cài hé dǎ lánqiú. Nǐ xiǎng bu xiǎng yìqǐ qù?

张爱华: 我想看电影和吃中国菜,但是我不想打篮球。

Zhāng Àihuá: Wǒ xiǎng kàn diànyǐng hé chī Zhōngguó cài, dànshì wǒ bù xiǎng dǎ lánqiú.

李小明: 没关系。我们明天不打篮球。我们听音乐怎么样?

Lǐ Xiǎomíng: Méi guānxi. Wǒmen míngtiān bù dǎ lánqiú. Wǒmen tīng yīnyuè zěnmeyàng?

张爱华: 行。我想听音乐。

Zhāng Àihuá: Xíng. Wǒ xiǎng tīng yīnyuè.

李小明：太好了！明天见。

Lǐ Xiǎomíng: Tài hǎo le! Míngtiān jiàn.

张爱华：明天见。

Zhāng Àihuá: Míngtiān jiàn.

Comprehension Questions

1. 张爱华明天想打篮球吗？
2. 李小明和张爱华明天想做什么？

情景(二) Scenario 2

王文丽：明天我想游泳。你想不想一起去？

Wáng Wénlì: Míngtiān wǒ xiǎng yóu yǒng. Nǐ xiǎng bu xiǎng yìqǐ qù?

吴家玲：明天我没有空儿。今天怎么样？

Wú Jiālíng: Míngtiān wǒ méiyǒu kòngr. Jīntiān zěnmeyàng?

王文丽：今天我没有空儿。后天呢？

Wáng Wénlì: Jīntiān wǒ méiyǒu kòngr. Hòutiān ne?

吴家玲：行！后天我有空儿。我们后天去吧。

Wú Jiālíng: Xíng! Hòutiān wǒ yǒu kòngr. Wǒmen hòutiān qù ba.

王文丽：好，再见。

Wáng Wénlì: Hǎo, zàijiàn.

吴家玲：再见。

Wú Jiālíng: Zàijiàn.

Comprehension Questions

1. 吴家玲明天有空儿吗？
2. 王文丽和吴家玲想今天去游泳吗？

口语沟通活动 Oral Communication Activities

I. 情况：下面列出了明天王文丽和吴家玲各自想做的四个活动。但是她们想一起做活动，所以得一起决定明天要做什么。

Situation: Wang Wenli and Wu Jialing each have four activities listed below that they feel like doing tomorrow. However, they want to do activities together and need to come to agreement on which activities they choose to do.

任务：两人一组。一个人扮演王文丽，另一个人扮演吴家玲。用下面列出的会话结构跟你的伙伴商量你们明天要一起做什么活动。

Task: Find a partner in class. One acts as Wang Wenli and the other acts as Wu Jialing. Use the conversation structure listed below to negotiate with your partner on the activity you will do together.

会话结构：

Conversation Structure:

王文丽：明天你想做什么？

吴家玲：明天我想游泳。你想不想一起去？

王文丽：我不想游泳。我想听音乐。你想不想一起去？

吴家玲：好。

王文丽想……
活动1：听音乐
活动2：吃中国菜
活动3：打篮球
活动4：看电影

吴家玲想……
活动1：游泳
活动2：吃法国菜
活动3：下棋
活动4：看电影

用上面句型问对方要做什么活动，直到你们找到一个两人明天都想做的活动。
Continue using the conversation structure above to invite each other to do each of the activities until you find an activity both of you want to do tomorrow.

II. 情况：今天是暑假第一天。陈林想邀他的朋友林丽一起看电影，但是陈林不知道林丽有没有空。
Situation: Today is the first day of summer vacation. Chen Lin wants to invite his friend, Lin Li to watch a movie together, but Chen Lin doesn't know when Lin Li is available.

任务：两人一组。一个人扮演陈林，另一个人扮演林丽。用下面陈林和林丽的日程表和列出的会话结构商量你们想什么时候去看电影。问问对方今天、明天和后天有没有空。
Task: Find a partner in class. One acts as Chen Lin and the other acts as Lin Li. Use Chen Lin's and Lin Li's schedules and the conversation structure listed below to come to an agreement on the day you two would like to watch a movie together. Ask about each other's availability for today, tomorrow and the day after tomorrow.

会话结构：
Conversation Structure:
陈林：今天我想看电影。你想不想一起去？
林丽：今天我没有空儿。明天怎么样？

林丽的日程表 林丽's schedule

今天	明天	后天
没有空儿	有空儿	有空儿

陈林的日程表 陈林's schedule

今天	明天	后天
有空儿	没有空儿	有空儿

用上面句型问对方什么时候有空儿,直到你们找到两人都有空儿的一天。

Continue using the conversation structure to ask for each other's availability until you find a day both of you can go to a movie together.

读写沟通活动 Literacy Communication Activities

I. 问你的同学们明天想做什么。在教室走动并问你的同学:"明天你想做什么?"。把他们的名字和答案写在下面空格里。

Investigate what activities your classmates feel like doing tomorrow. Walk around the classroom and find out the answers by asking your classmates, "明天你想做什么?". Write down their names and all the answers you get.

姓名	明天你想做什么?

II. 你的朋友在你的脸书写了短信邀你一起出去。在下面回他们的信息。

Your friends posted messages on your Facebook page to invite you to do activities together. Reply to their messages by writing your answers in the blanks.

1. 明天我想去吃中国菜。你想不想一起去？

吴家玲

____

你的姓名：Your name:

____

2. 今天你想做什么？我想去学中文。你想不想一起去？

王文丽

____

你的姓名：Your name:

____

3. 后天你有空儿吗？后天我想去游泳。你想不想一起去？

张爱华

____

你的姓名：Your name:

____

讨论 Discussion

Activities Chinese People Like to do

Do you know what activities many Chinese people like to do for leisure? Take a look at the pictures below and discuss with your instructor and classmates what activities you think Chinese people like to do.

泡茶

打麻将

逛夜市

做早操

唱卡拉OK

下棋

Discussion Question:

Are the activities Chinese people do different from Americans?

第七课 Lesson 7

长大以后你想做什么？
What Do You Want to Be When You Grow Up?

沟通任务 Communication Tasks

- *Asking what someone wants to be when he or she grows up*
- *Wishing for someone's dreams to come true*
- *Asking why someone wants to have a certain type of job when he or she grows up*

课前讨论 Warm-up Discussion Questions

1. What do you want to be when you grow up?
2. Is there a certain type of job that you would like to do in the future? What are the reasons that you would like to do it?

生词 Vocabulary

1	长大	長大	zhǎngdà	to grow up

◎ 我想长大。(I want to grow up.)

2 以后	以後	yǐhòu	after	*n.*
		◎ 长大以后我想去中国。(I want to go to China after I grow up.)		
3 医生	醫生	yīshēng	medical doctor	*n.*
		◎ 我想做医生。(I want to be a doctor.)		
4 太棒了	太棒了	tài bàng le	great	
		◎ 太棒了。(Great.)		
5 祝	祝	zhù	to wish	*v.*
		◎ 祝好。(I wish you a good day.)		
6 你们	你們	nǐmen	you (plural)	*pron.*
		◎ 你们都是中文老师吗?(Are you all Chinese teachers?)		
7 梦想成真	夢想成真	mèngxiǎng chéng zhēn	dreams come true	
		◎ 祝你梦想成真。(I wish for your dreams to come true.)		
梦想	夢想	mèngxiǎng	dream	*n.*
成真	成真	chéng zhēn	to come true	
8 画家	畫家	huàjiā	painter	*n.*
		◎ 我爸爸是画家。(My father is a painter.)		
9 画画儿	畫畫兒	huà huàr	to paint	*v.o.*
		◎ 我爸爸喜欢画画。(My father likes to paint.)		
10 作家	作家	zuòjiā	writer	*n.*
		◎ 我想做作家。(I want to be a writer.)		
11 写作	寫作	xiězuò	to write	*v.*
		◎ 我喜欢写作。(I like to write.)		

补充词语 Supplementary Vocabulary

1	警察	警察	jǐngchá	police	*n.*
2	运动员	運動員	yùndòngyuán	athlete	*n.*
3	书法家	書法家	shūfǎjiā	calligrapher	*n.*
4	工程师	工程師	gōngchéngshī	engineer	*n.*
5	音乐家	音樂家	yīnyuèjiā	musician	*n.*
6	舞蹈家	舞蹈家	wǔdǎojiā	dancer	*n.*
7	运动	運動	yùndòng	to exercise	*v.*
8	书法	書法	shūfǎ	calligraphy	*n.*
9	跳舞	跳舞	tiào wǔ	to dance	*v.o.*
10	教书	教書	jiāo shū	to teach	*v.o.*

沟通任务 Communication Tasks

I Asking what someone wants to be when he or she grows up

Question: 长大以后你想做什么？(What do you want to be when you grow up?)

Answer: 长大以后我想做________。(I want to be ________ when I grow up.)

Example:

Question: 长大以后你想做什么？(What do you want to be when you grow up?)

Answer: 长大以后我想做医生。(I want to be a doctor when I grow up.)

Practice: Find a partner. Tell him/her that you want to be a teacher when you grow up and ask them what he or she wants to be when he or she grows up.

II Wishing for someone's dreams to come true

祝________梦想成真！(I wish ________ dreams come true!)

Example:

祝你梦想成真!(I wish for your dreams to come true!)

祝王老师梦想成真!(I wish for teacher Wang's dreams to come true!)

Practice: Tell your classmates sitting next to you "I wish for your dreams to come true!".

III Asking why someone wants to have a certain type of job when he or she grows up

Question:为什么你想做________?(Why do you want to be ________?)

Answer:因为我很喜欢________。(Because I like ________ very much.)

Examples:

Question:为什么你想做画家?(Why do you want to be a painter?)

Answer:因为我很喜欢画画儿。(Because I like to paint very much.)

Practice: Work in pairs. Without reading the example above, ask your partner, "why do you want to be a painter?" Your partner needs to respond by saying "Because I like to paint very much."

情景(一) Scenario 1

王老师: 李小明,长大以后你想做什么?

Wáng Lǎoshī: Lǐ Xiǎomíng, zhǎngdà yǐhòu nǐ xiǎng zuò shénme?

李小明: 长大以后我想做医生。

Lǐ Xiǎomíng: Zhǎngdà yǐhòu wǒ xiǎng zuò yīshēng.

王老师: 张爱华,长大以后你想做什么?

Wáng Lǎoshī: Zhāng Àihuá, zhǎngdà yǐhòu nǐ xiǎng zuò shénme?

张爱华: 我想做中文老师。

Zhāng Àihuá: Wǒ xiǎng zuò Zhōngwén lǎoshī.

王老师: 太棒了! 祝你们梦想成真!

Wáng Lǎoshī: Tài bàng le! Zhù nǐmen mèngxiǎng chéng zhēn!

Comprehension Questions

1. 李小明长大以后想做什么？
2. 张爱华长大以后想做什么？

情景（二）Scenario 2

王文丽：长大以后我想做画家。
Wáng Wénlì: Zhǎngdà yǐhòu wǒ xiǎng zuò huàjiā.
吴家玲：为什么你想做画家？
Wú Jiālíng: Wèi shénme nǐ xiǎng zuò huàjiā?
王文丽：因为我很喜欢画画儿。你呢？长大以后你想做什么？
Wáng Wénlì: Yīnwèi wǒ hěn xǐhuan huà huar. Nǐ ne? Zhǎngdà yǐhòu nǐ xiǎng zuò shénme?
吴家玲：我想做作家，因为我很喜欢写作。
Wú Jiālíng: Wǒ xiǎng zuò zuòjiā, yīnwèi wǒ hěn xǐhuan xiězuò.

Comprehension Questions

1. 为什么王文丽想做画家？
2. 为什么吴家玲想做作家？

口语沟通活动 Oral Communication Activities

I. 情况：下面五张图片显示了李中、张丽、王强、陈朋和陈龙长大以后想做什么。

Situation: The five pictures illustrated below show what Li Zhong, Zhang Li, Wang Qiang, Chen Peng, and Chen Long want to be when they grow up.

图一 李中

图二 张丽

图三 王强

图四 陈朋

图五 陈龙

任务：两人一组。用下面列出的新词和会话结构轮流问对方有关上面五个图的问题。

Task: Work with a partner and take turns to ask each other questions about the five pictures using the new vocabulary and conversation structure shown below.

职业 Jobs
1. 警察
2. 运动员
3. 书法家
4. 工程师
5. 医生

会话结构：

Conversation Structure:

A: 长大以后李中想做什么？

B: 长大以后李中想做警察。

II. 情况：在做课前讨论时，你听见你的同学用英文说他们长大以后想做什么。

Situation: In the warm-up discussion, you learned in English what your classmates want to be when they grow up.

任务：用中文和下面的会话结构问至少五个同学他们长大以后想做什么。在下面表格记录他们的答案。结束对话前祝他们梦想成真。

Task: Ask at least five classmates in Chinese what they want to be when they grow up using the conversation structure below and write down their answers in the chart below. End your conversation with them by wishing for their dreams to come true.

会话结构：

Conversation Structure:

你：长大以后你想做什么？

你同学：我想做________________。

你：祝你梦想成真！

名字	长大以后我想……
1.	
2.	
3.	
4.	
5.	

读写沟通活动 Literacy Communication Activities

I. 你的老师会请志愿者演出下表内列出的职业。当志愿者演出一个职业时，大家要猜志愿者演出的职业是什么并说："长大以后你想做__________，因为你很喜欢__________。"

Your teacher will call volunteers to act out each of the jobs listed below. When a volunteer acts out a job, the class will guess what job the volunteer is acting out by saying: "长大以后你想做__________，因为你很喜欢__________。"

职业 Jobs	喜欢职业的原因 Possible reasons one likes the jobs
职业1.Job 1. 运动员	很喜欢运动
职业2.Job 2. 书法家	很喜欢书法
职业3.Job 3. 音乐家	很喜欢音乐
职业4.Job 4. 舞蹈家	很喜欢跳舞
职业5.Job 5. 老师	很喜欢教书

例子:职业 1 (Example: Job 1)
长大以后你想做<u>运动员</u>,因为你很喜欢<u>运动</u>。

志愿者会告诉你你猜对了还是错了。如果对了,把志愿者的姓名和演出的职业写在下表内。例子见第一行。

The volunteers will tell you whether or not your guesses are correct. If your guesses are correct, record the volunteers' names and the jobs they act out in the table below. See the example in the first row.

姓名 Names	职业 Jobs
李小明	老师

II. 你的朋友林友在你的脸书，写了一个短信，问你长大以后想做什么。在下面回复他的短信。

Your friend Lin You posted a message on your Facebook page to ask you what you want to be when you grow up. Reply to his message by writing your answers in the blanks.

1. 

长大以后你想做什么?

林友

____

你的姓名:Your name:

____

讨论Discussion

Careers in China

With the economic growth of China and its increasing international political influence, China has become an attractive market for business. According to the Ministry of Labor and Social Security, China has been creating millions of new jobs every year.

The most profitable careers in China are (China Service Mall, 2011):

Top 1 Salesman

Top 2 Computer Software Developer

Top 3 Building Contractor

Top 4 Security Broker

Top 5 Certified Public Accountant

Top 6 Advertiser

Top 7 Special Culturist

Top 8 Lawyer

Top 9 Public Relations

Top 10 Orthopedic Doctor and Beautician

Discussion Questions:

1. Do you plan to work in China when you grow up?
2. What is your dream job in China? Is it in the top 10 list?
3. Ask your teacher what he or she knows about the popular careers in China.

Reference:

China Service Mall (2011). Retrieved on July 20, 2011, from http://www.at0086.com/at_News/319/577918949059123.html

第八课 Lesson 8

你想喝什么？
What Would You Like to Drink?

沟通任务 Communication Tasks

- *Welcoming people to your home*
- *Asking what someone wants to drink*
- *Asking what drinks are available*

课前讨论 Warm-up Discussion Questions

1. What is your favorite drink?
2. What do you say to your friends when they enter the door to visit your home?

生词 Vocabulary

1 来	來	lái	to come	*v.*
		◎ 来我家。(Come to my home.)		
2 欢迎	歡迎	huānyíng	to welcome	*v.*
		◎ 欢迎来我家。(Welcome to my home.)		

3 喝	喝	hē	to drink	*v.*
		◎ 你想喝什么？(What would you like to drink?)		
4 可乐	可樂	kělè	coke	*n.*
		◎ 我想喝可乐。(I would like to drink coke.)		
5 水	水	shuǐ	water	*n.*
		◎ 我想喝水。(I would like to drink water.)		
6 冰茶	冰茶	bīngchá	iced tea	*n.*
		◎ 我想喝冰茶。(I would like to drink iced tea.)		
7 爷爷	爺爺	yéye	grandfather	*n.*
		◎ 我爷爷想喝茶。(My grandfather would like to drink tea.)		
8 奶奶	奶奶	nǎinai	grandmother	*n.*
		◎ 我奶奶也想喝茶。(My grandmother also would like to drink tea.)		
9 咖啡	咖啡	kāfēi	coffee	*n.*
		◎ 他妈妈想喝咖啡。(His mother would like to drink coffee.)		
10 红茶	紅茶	hóngchá	black tea	*n.*
		◎ 你爸爸想喝红茶。(Your father would like to drink black tea.)		
11 果汁	果汁	guǒzhī	juice	*n.*
		◎ 我弟弟想喝果汁。(My younger brother would like to drink juice.)		
12 饮料	飲料	yǐnliào	drink	*n.*
		◎ 你有什么饮料？(What drinks do you have?)		
13 谢谢	謝謝	xièxie	to thank	*v.*
		◎ 谢谢你。(Thank you.)		

补充词语 Supplementary Vocabulary

1 绿茶	綠茶	lǜchá	green tea	*n.*
2 汽水	汽水	qìshuǐ	soda	*n.*
3 水果茶	水果茶	shuǐguǒchá	fruit tea	*n.*
4 牛奶	牛奶	niúnǎi	milk	*n.*
5 沙士	沙士	shāshì	root beer	*n.*

沟通任务 Communication Tasks

I Welcoming people to your home

欢迎来__________！(Welcome to __________!)

Examples:

欢迎来我家！(Welcome to my home!)

欢迎来我们家！(Welcome to our home!)

Practice: Act out being at your home and your classmates come to visit you. Tell them "欢迎来我家!" when you open the door for them.

II Asking what someone wants to drink

Question: 你想喝什么？(What would you like to drink?)

Answer: 我想喝__________。谢谢！(I would like to drink __________. Thanks!)

Attention:

1. 你 vs. 您: 您 is the polite form of 你. 你 is used with close friends or acquaintances. 您 is used with elders or people of higher rank in an organization.

Example:Using 你

Question: 你想喝什么？(What would you like to drink?)

Answer: 我想喝可乐。谢谢！(I would like to drink coke. Thanks!)

Example: Using 您

Question: 爷爷，请问您想喝什么？(What would you like to drink, grandfather?)

Answer: 我想喝咖啡。谢谢！(I would like to drink coffee. Thanks!)

Practice: Ask your teacher what he or she would like to drink. Use 您 in the question.

III Asking what drinks are available

Question: 你有什么饮料？(What drinks do you have?)

Answer: 我有__________________。(I have __________________.)

Example:

Question: 你有什么饮料？(What drinks do you have?)

Answer: 我有咖啡、红茶、可乐和果汁。(I have coffee, black tea, coke and juice.)

Practice: Work with a partner. Ask each other what drinks they have.

情景(一) Scenario 1

李小明邀请张爱华、王文丽和吴家玲到他家开派对。

李小明： 欢迎来我家！你们想喝什么？

Lǐ Xiǎomíng：Huānyíng lái wǒ jiā! Nǐmen xiǎng hē shénme?

王文丽： 我想喝可乐。

Wáng Wénlì: Wǒ xiǎng hē kělè.

张爱华： 我想喝水。

Zhāng Àihuá: Wǒ xiǎng hē shuǐ.

吴家玲： 我想喝冰茶。

Wú Jiālíng: Wǒ xiǎng hē bīngchá.

李小明拿饮料给他的朋友们喝。

王文丽、张爱华和吴家玲： 谢谢！

Wáng Wénlì, Zhāng Àihuá hé Wú Jiālíng: Xièxie!

Comprehension Questions

1. 王文丽想喝什么？
2. 张爱华想喝什么？
3. 吴家玲想喝什么？

情景(二) Scenario 2

吴家玲的爷爷奶奶来吴家玲的家看她。

吴家玲： 爷爷奶奶，欢迎来我家！请问您想喝什么？

Wú Jiālíng: Yéye nǎinai, huānyíng lái wǒ jiā! Qǐngwèn nín xiǎng hē shénme?

爷爷： 你有什么饮料？

Yéye: Nǐ yǒu shénme yǐnliào?

吴家玲： 我有咖啡、红茶、可乐和果汁。

Wú Jiālíng: Wǒ yǒu kāfēi, hóngchá, kělè hé guǒzhī.

爷爷： 我想喝咖啡。

Yéye: Wǒ xiǎng hē kāfēi.

奶奶： 我想喝红茶。

Nǎinai: Wǒ xiǎng hē hóngchá.

Comprehension Questions

1. 吴家玲有什么饮料？
2. 爷爷想喝什么？
3. 奶奶想喝什么？

口语沟通活动 Oral Communication Activities

I. 情况：王老师去看他的学生林强，林强欢迎王老师来看他并请他喝饮料。

Situation: Wang Laoshi goes to visit his student, Lin Qiang. Lin Qiang welcomes Wang Laoshi and offers him a drink.

任务：两人一组。一个人扮演王老师，另一个人扮演林强。用下面的会话结构做角色扮演。

Task: Work in pairs. One acts as Wang Laoshi and the other acts as Lin Qiang. Use the following conversation structure to act out the home visit.

会话结构：

Conversation Structure:

林强：王老师，欢迎来我家！请问您想喝什么？

王老师：我想喝______________。谢谢！

II. 情况：下表列出的是一家饮料店的饮料单。服务员会问客人他们想喝什么饮料，并在每个客人点的饮料名字旁边打钩。

Situation: The table listed below is the drinks list of a drink store. The waiter or waitress will ask the customers what they would like to drink and will put a check mark on the list next to the drink that each customer orders.

饮料	
□ 冰茶	□ 汽水
□ 可乐	□ 咖啡
□ 红茶	□ 沙士
□ 绿茶	□ 果汁
□ 水果茶	□ 牛奶

任务:小组练习。一个人扮演服务员,其他人扮演客人。用下面列出的会话结构做角色扮演。最后服务员得在班上报告他的客人点了什么。

Task: Work in small groups. One student acts as the waiter or waitress and the rest of the students act as the customers. Use the following conversation structure to act out the scene. The waiter/waitress will report to the class what his or her customers ordered.

会话结构:

Conversation Structure:

服务员 Waiter/Waitress:请问您想喝什么?

客人1 Customer 1:你有什么饮料?

服务员 Waiter/Waitress:我有________________。

客人1 Customer 1:我想喝________________。

客人2 Customer 2:我想喝________________。

客人3 Customer 3:我想喝________________。

读写沟通活动 Literacy Communication Activities

I. 在一张纸条上写你想喝什么饮料。把纸条折好交给老师。你的老师会把纸条打开，请同学们大声读纸条上的句子。班上的同学得记录几个同学喜欢什么饮料。

Write a note which lists what you would like to drink. Fold the note and submit it to your teacher. Your teacher will unfold each note and ask the class to read what is written in the note. The class will tally how many students like each type of drink.

例子：Example:

我想喝可乐。

II. 下图里的家人想喝不同的饮料。在图片下面写出他们每个人想喝什么。

The family members in the pictures below would like to drink different kinds of beverages. Write what each of them wants to drink in the line below each picture.

图一 爸爸

爸爸想喝________________。

图二 妈妈

妈妈想喝________________。

图三 弟弟

弟弟想喝______________。

图四 妹妹

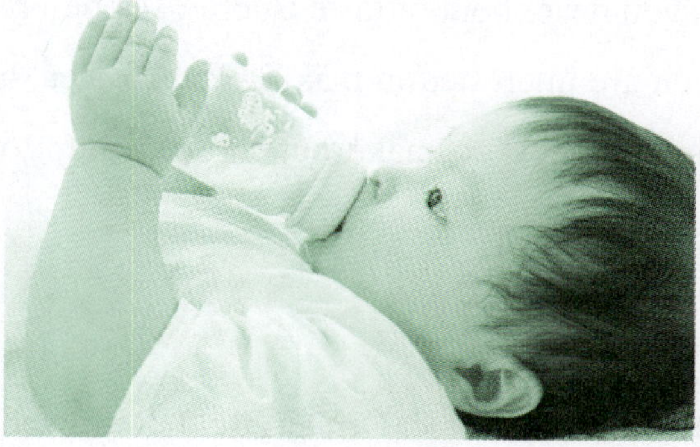

妹妹想喝______________。

讨论 Discussion

Bubble Tea

Bubble tea (珍珠奶茶 zhēnzhū nǎichá) is a sweet and cold tea beverage invented in Taiwan in the 1980s. Bubble tea is also known as pearl tea, boba milk tea, and tapioca milk tea. The basic ingredients of Bubble tea include tea, milk, and tapioca balls.

Bubble tea has become world famous as Bubble tea shops are available all over the world. For instance, there are hundreds of Bubble tea shops in many Asian countries, Canada and the United States. In the United Kingdom the first Bubble tea shop was opened in April 2011 by a British investment banker, Assad Khan, who first tasted Bubble tea in New York few years prior. Khan estimated selling 250 cups of tea a day in his shop but his sales numbers are actually 500 cups per day. Bubble tea is so popular that Khan decided to open a second shop in July 2011 in a department store in London.

Discussion Questions:

1. Did you hear about or taste Bubble Tea before? Do you like the drink?
2. If you are interested in making Bubble Tea, have a Bubble Tea party in class or at home. Here is a link to "how to make bubble tea": http://www.bubbleteasupply.com/index.php?page=recipes.html

Reference:

Bubble Tea: The Perfect Drink (2011). Retrieved on July 22, 2011, from http://www.bubbleteasupply.com/index.php?page=what.html

第九课 Lesson 9

今天天气怎么样？How is the Weather Today?

沟通任务 Communication Tasks

- *Asking how the weather is today*
- *Asking what the weather forecast says about the weather tomorrow*
- *Describing the weather*
- *Stating what you can or are willing to do in certain situations/conditions*
- *Reminding someone not to forget to do something*

课前讨论 Warm-up Discussion Questions

1. What kind of weather do you like? Do you like sunny weather or rainy weather?
2. Do you check the weather forecast regularly?

生词 Vocabulary

1 天气	天氣	tiānqì	weather	*n.*

◎ 天气很好。(The weather is nice.)

2 上午	上午	shàngwǔ	morning	*n.*
		◎ 上午天气很好。 (The weather is nice in the morning.)		
3 下午	下午	xiàwǔ	afternoon	*n.*
		◎ 今天下午天气不好。 (The weather is not nice this afternoon.)		
4 有点儿	有點兒	yǒudiǎnr	a bit	*adv.*
		◎ 今天早上有点儿冷。(It is a bit chilly this morning.)		
5 冷	冷	lěng	cold, chilly	*adj.*
		◎ 今天下午有点儿冷。(It is a bit chilly this afternoon.)		
6 天气预报	天氣預報	tiānqì yùbào	weather forecast	
		◎ 天气预报说今天早上天气很好。 (The weather forecast says the weather is nice this morning.)		
7 晚上	晚上	wǎnshang	evening	*n.*
		◎ 今天晚上不冷。(It is not cold this evening.)		
8 会	會	huì	will	*aux.*
		◎ 今天晚上会下雨。(It will rain tonight.)		
9 下雪	下雪	xià xuě	to snow	
		◎ 今天晚上会下雪。(It will snow tonight.)		
10 可以	可以	kěyǐ	can	*aux.*
		◎ 我可以喝茶。(I can drink tea.)		
11 雪人	雪人	xuěrén	snowman	*n.*
		◎ 你喜欢雪人吗? (Do you like the snowman?)		
12 下雨	下雨	xià yǔ	to rain	*v.o.*
		◎ 明天晚上会下雨。(It will rain tomorrow evening.)		
13 出太阳	出太陽	chū tàiyáng	the sun comes out	*v.o.*
		◎ 今天下午会出太阳。 (The sun will come out this afternoon.)		

14 所以	所以	suǒyǐ	therefore, so	conj.
		◎ 今天会下雪，所以我可以做雪人。 (It will snow tomorrow, so I can make a snowman.)		
15 下雨天	下雨天	xià yǔ tiān	rainy day	
		◎ 今天是下雨天。(Today is a rainy day.)		
16 晴天	晴天	qíngtiān	sunny day	n.
		◎ 今天是晴天。(Today is a sunny day.)		
17 别忘了	别忘了	bié wàng le	do not forget	
		◎ 别忘了带你朋友来学中文。 (Do not forget to bring your friends to study Chinese.)		
18 带	帶	dài	to bring	v.
		◎ 别忘了带伞。(Do not forget to bring an umbrella.)		
19 伞	傘	sǎn	umbrella	n.
		◎ 今天是下雨天，别忘了带伞。 (Today is a rainy day, do not forget to bring an umbrella.)		

补充词语 Supplementary Vocabulary

1 热	熱	rè	hot	adj.
2 阴天	陰天	yīntiān	cloudy day	n.
3 昨天	昨天	zuótiān	yesterday	n.
4 前天	前天	qiántiān	the day before yesterday	n.

沟通任务 Communication Tasks

I Asking how the weather is today

Question：今天天气怎么样？(How is the weather today?)

Answer：今天天气____________。(The weather today is ____________.)

Example:

Question: 今天天气怎么样？(How is the weather today?)

Answer 1: 今天天气很好。(The weather today is very nice.)

Answer 2: 今天天气有点儿冷。(The weather today is a bit chilly.)

Answer 3: 今天天气有点儿热。(The weather today is a bit hot.)

Answer 4: 今天上午天气很好，但是下午有点儿冷。

(The weather was very nice this morning, but a bit chilly this afternoon.)

Practice: Ask a classmate how the weather is today. Is the weather nice, cold, or hot?

II Asking what the weather forecast says about the weather tomorrow

Question: 天气预报说明天天气怎么样？

(What does the weather forecast say about the weather tomorrow?)

Answer: 天气预报说明天______________。

(The weather forecast says that ______________ tomorrow.)

Example:

Question: 天气预报说明天天气怎么样？

(What does the weather forecast say about the weather tomorrow?)

Answer 1: 天气预报说明天会下雨。

(The weather forecast says that it will rain tomorrow.)

Answer 2: 天气预报说明天上午会下雨、下午会出太阳。

(The weather forecast says that it will rain tomorrow morning, but the sun will come out in the afternoon.)

Practice: Your classmate sitting next to you is going to ask you what the weather forecast says about the weather tomorrow. Tell your classmate that it will snow tomorrow.

III Describing the weather

明天是______________天。(Tomorrow will be a ______________ day.)

Examples:

明天是下雨天。(Tomorrow will be a rainy day.)

明天是晴天。(Tomorrow will be a sunny day.)

明天是阴天。(Tomorrow will be a cloudy day.)

Practice: Predict what the weather will be tomorrow and tell the prediction to your class.

IV Stating what you can or are willing to do in certain situations/conditions

我可以______________。(I can ______________.)

Examples:

Condition: It will snow tomorrow.

Statement: 我可以做雪人。(I can make a snowman.)

Situation: Your friend asks you what you can bring to the party tomorrow.

Statement: 我可以带饮料。(I can bring drinks.)

Practice: Your teacher has a meeting and needs to step out of the class for a few minutes. He or she wants to know what you can do without him or her in class. Tell your teacher that you can practice writing Chinese characters.

V Reminding someone not to forget to do something

别忘了______________。

(Don't forget to ______________.)

Examples:

别忘了带伞。(Don't forget to bring an umbrella.)

别忘了练习说中文。(Don't forget to practice speaking Chinese.)

Practice: Your classmate sitting next to you promised you that he or she would show you his or her Chinese books tomorrow. Remind your classmate not to forget to bring his or her Chinese books tomorrow.

情景(一) Scenario 1

李小明刚从学校回到家。

妈妈: 今天天气怎么样?

Māma: Jīntiān tiānqì zěnmeyàng?

李小明: 今天上午天气很好,但是下午有点儿冷。

Lǐ Xiǎomíng: Jīntiān shàngwǔ tiānqì hěnhǎo, dànshì xiàwǔ yǒudiǎnr lěng.

妈妈: 天气预报说今天晚上会下雪。

Māma: Tiānqì yùbào shuō jīntiān wǎnshang huì xià xuě.

李小明: 太棒了! 我可以做雪人。

Lǐ Xiǎomíng: Tài bàng le! Wǒ kěyǐ zuò xuěrén.

Comprehension Questions

1. 今天上午天气怎么样?
2. 今天下午天气怎么样?
3. 李小明想做什么?

情景(二) Scenario 2

爷爷刚看完天气预报。

吴家玲: 天气预报说明天天气怎么样?

Wú Jiālíng: Tiānqì yùbào shuō míngtiān tiānqì zěnmeyàng?

爷爷: 天气预报说明天上午会下雨、下午会出太阳。

Yéye: Tiānqì yùbào shuō míngtiān shàngwǔ huì xià yǔ, xiàwǔ huì chū tàiyáng.

吴家玲: 所以明天上午是下雨天,下午是晴天。

Wú Jiālíng: Suǒyǐ míngtiān shàngwǔ shì xià yǔ tiān, xiàwǔ shì qíngtiān.

爷爷: 是啊,明天上午别忘了带伞。

Yéye: Shì a, míngtiān shàngwǔ bié wàng le dài sǎn.

Comprehension Questions

1. 天气预报说明天天气怎么样?
2. 爷爷说明天上午别忘了带什么?

口语沟通活动 Oral Communication Activities

I. 情况:你的老师准备了一些纸条。纸条上写了一些天气预报的句子。

Situation: Your teacher has prepared folded paper slips with the following weather forecast statements.

明天会下雨。

明天会出太阳。

明天会下雪。

明天有点儿热。

明天有点儿冷。

任务：你的老师会一次请一个志愿者打开一张纸条并表演纸条上的天气预报。演出后，志愿者会问大家："明天天气怎么样？"大家得猜一猜志愿者表演了上面哪个句子。

Task: Your teacher will call volunteers one at a time to unfold one of the paper slips and act out the statement written in the paper slip. After a volunteer acts out the statement, he or she will ask the class "明天天气怎么样?" and the class will guess which statement the volunteer has acted out (See the conversation structure below).

会话结构：

Conversation Structure:

（志愿者表演后）(After a volunteer acts out the statement)

志愿者 Volunteer：明天天气怎么样？

大家 Class：明天天气____________。

II. 情况：气象预报员为明天的天气做了以下预报。

Situation: The weather man/woman reported the following weather statement about the weather tomorrow.

"明天上午天气不好，是阴天。明天下午会下雨。明天晚上有点儿冷。"

任务：你的老师扮演气象预报员。他会大声朗读上方的天气预报。只听，但不读他的天气预报。听完后，两人一组用下面的会话结构问对方明天上午、下午和晚上的天气怎么样。

Task: Your teacher will act as the weather man/woman and read aloud the weather statement above. Listen to the weather man/woman without reading the written statement above. When the weather man/woman finishes reporting about the weather, work in pairs to ask each other the weather forecast for tomorrow morning, afternoon, and evening using the conversation structure listed below.

会话结构：
Conversation Structure:
(明天上午的天气预报 weather forecast for tomorrow morning)
Student A: 天气预报说明天上午天气怎么样?
Student B: 天气预报说明天上午______________。

(明天下午的天气预报 weather forecast for tomorrow afternoon)
Student B: 天气预报说明天下午天气怎么样?
Student A: 天气预报说明天下午______________。

(明天晚上的天气预报 weather forecast for tomorrow evening)
Student B: 天气预报说明天晚上天气怎么样?
Student A: 天气预报说明天晚上______________。

读写沟通活动 Literacy Communication Activities

I. 你和你的朋友在 *QQ* 上聊明天要带什么到你的生日派对。你们的对话已经用英文写好了。两人一组把对话翻译成中文。

Your friends and you are having a conversation on QQ about what to bring to your birthday party tomorrow. The conversation is already written in English. Work with a classmate and translate the conversation into Chinese.

你

朋友 1: 我可以 ______________________

(bring coke)

朋友 2: ______________________

(I can bring iced tea.)

你: ______________________

(Don't forget to bring an umbrella. It will rain tomorrow!)

II. 下图显示了前天、昨天、今天、明天和后天的天气。在每个图的下面描述图里的天气。

The pictures below show the weather for the day before yesterday, yesterday, today, tomorrow, and the day after tomorrow. Write the description of the weather in the line below each picture.

图一 前天

前天是________天。

图二 昨天

昨天是________天。

图三 今天

今天是________天。

图四 明天

明天________。

图五 后天

后天是________天。

讨论 Discussion

The Habit of Carrying an Umbrella

Western people like to get a tan because browning their skin makes them look healthier. However, many Asian groups, including Chinese and especially women, go to great lengths to protect their skin from the sun. It is common to see Chinese women using umbrellas or parasols or wearing long sleeves to hide from the sun on a sunny day. The reason Chinese women don't like getting tan is because of the standards of beauty in Asia. It is considered unattractive to have freckles and sun spots. On the other hand, having pale and flawless skin is considered beautiful. Therefore, Chinese women use many means to keep their skin light. In addition to using umbrellas and wearing long sleeves, sun block and skin whitening products are popular in skin care stores.

Discussion Question:

With your teacher's help, compare and contrast the standards of beauty in your culture vs. Chinese culture. What are the differences and the similarities?

第十课
Lesson 10

早餐吃了吗？
Have You Eaten Breakfast Yet?

沟通任务 Communication Tasks

- *Knowing how to greet someone at different times of a day*
- *Making a command*
- *Describing what there is for breakfast/lunch/dinner*
- *Asking how someone is*
- *Telling someone you need to eat something*
- *Telling people what you have to do*
- *Asking if someone has eaten something yet*

课前讨论 Warm-up Discussion Questions

1. Do you usually eat breakfast before going to school? What do you usually have for breakfast?
2. Have you had an authentic Chinese meal? What did you eat?

生词 Vocabulary

1 早安	早安	zǎo'ān	good morning	
早上好	早上好	zǎoshang hǎo		
		◎ 早安/早上好。(Good morning.)		
2 早餐	早餐	zǎocān	breakfast	*n.*
		◎ 我想吃早餐。(I want to eat breakfast.)		
3 蛋	蛋	dàn	egg	*n.*
		◎ 我喜欢吃蛋。(I like to eat eggs.)		
4 吐司	吐司	tǔsī	toast	*n.*
		◎ 我不喜欢吃吐司。(I don't like to eat toast.)		
5 香肠	香腸	xiāngcháng	sausage	*n.*
		◎ 我喜欢吃香肠。(I like to eat sausage.)		
6 牛奶	牛奶	niúnǎi	milk	*n.*
		◎ 我想喝牛奶。(I want to drink milk.)		
7 午安	午安	wǔ'ān	good afternoon	
下午好	下午好	xiàwǔ hǎo		
		◎ 午安/下午好。(Good afternoon.)		
8 饿	餓	è	hungry	*adj.*
		◎ 我很饿。(I am hungry.)		
9 午餐	午餐	wǔcān	lunch	*n.*
		◎ 我想吃午餐。(I want to eat lunch.)		
10 现在	現在	xiànzài	now	*n.*
		◎ 我现在不饿。(I am not hungry now.)		
11 得	得	děi	have to	*aux.*
		◎ 我现在得吃早餐。(I have to eat breakfast now.)		

12 上课	上課	shàng kè	to go to class	v.o.
		◎ 我现在得去上中文课。(I have to go to Chinese class now.)		
13 时间	時間	shíjiān	time	n.
		◎ 我现在没有时间。(I don't have time now.)		
14 吃了吗	吃了嗎	chī le ma	has eaten yet	
		◎ 你早餐吃了吗？(Have you eaten breakfast yet?)		
了	了	le	particle used to indicate completion of an action	part.
15 还	還	hái	still	adv.
		◎ 我还没吃。(I have not eaten yet.)		
16 晚餐	晚餐	wǎncān	dinner	n.
		◎ 我还没吃晚餐。(I have not eaten dinner yet.)		
17 晚上好	晚上好	wǎnshang hǎo	good evening	
		◎ 晚上好。(Good evening.)		
18 晚安	晚安	wǎn'ān	good night	
		◎ 晚安。(Good night.)		

补充词语 Supplementary Vocabulary

1 三明治	三明治	sānmíngzhì	sandwich	n.
2 汉堡	漢堡	hànbǎo	hamburger	n.
3 意大利面	意大利麵	Yìdàlì miàn	spaghetti	n.
4 炒饭	炒飯	chǎofàn	fried rice	n.
5 鸡肉	雞肉	jīròu	chicken	n.
6 青菜	青菜	qīngcài	leafy greens	n.
7 蛋花汤	蛋花湯	dànhuā tāng	egg drop soup	n.

沟通任务 Communication Tasks

I Knowing how to greet someone at different times of a day

Greeting in the morning: 早安/早上好。(Good morning.)

Greeting at noon or in the afternoon: 午安 /下午好。(Good afternoon.)

Greeting in the evening: 晚上好。(Good evening.)

Greeting at night: 晚安。(Good night.)

Practice: What time is it now? If it is before 12 pm, tell your class, "good morning" in Chinese. If it is after 12 pm, tell your class "good afternoon" in Chinese.

II Making a commend

来__________。(Come to __________.)

Examples:

来吃早餐。(Come to eat breakfast.)

来写汉字。(Come to write Chinese characters.)

Practice: Command the class to practice Chinese.

III Describing what there is for breakfast/lunch/dinner

早餐有__________________________。

(Breakfast includes __________________________.)

午餐有__________________________。

(Lunch includes __________________________.)

晚餐有__________________________。

(Dinner includes __________________________.)

Example:

早餐有蛋、吐司、香肠、和牛奶。

(Breakfast includes eggs, toast, sausage, and milk.)

Practice: What did you have for breakfast? Tell your classmates.

IV Asking how someone is

Question: 你好吗？(How are you?)

Positive Answer: 我很好。(I am great.)

Negative Answer: 我不太好。(I am not too good.)

Practice: Greet your classmate next to you.

V Telling someone you need to eat something

我很饿。我想吃__________。(I am very hungry. I want to eat __________.)

Example:

我很饿。我想吃<u>午餐</u>。(I am very hungry. I want to eat <u>lunch</u>.)

Practice: Find a partner and tell them that you are hungry and want to eat breakfast.

VI Telling people what you have to do

我现在得__________。(I have to __________ now.)

Give a more elaborate statement by telling people what you cannot do.

我现在得__________,没有时间__________。

(I have to __________ now, and don't have time to __________.)

Examples:

我现在得<u>去上课</u>。(I have to <u>go to a class</u> now.)

我现在得<u>练习说中文</u>。(I have to <u>practice speaking Chinese</u> now.)

我现在得<u>去上课</u>,没有时间<u>吃午餐</u>。

(I have to <u>go to a class</u> now, and don't have time to <u>eat lunch</u>.)

我现在得<u>练习说中文</u>,没有时间<u>画画儿</u>。

(I have to <u>practice speaking Chinese</u> now, and don't have time to <u>paint</u>.)

Practice: Tell your classmate sitting next to you that you have to go to a class and don't have time to practice Chinese.

VII Asking if someone has eaten something yet

Question: 你__________吃了吗？(Have you eaten __________ yet?)

Positive Answer: 吃了。(I have.)

Negative Answer: 还没吃。(I have not.)

Example:

Question: 你晚餐吃了吗？(Have you eaten dinner yet?)

Negative Answer: 还没吃。(I have not.)

Positive Answer: 吃了。(I have.)

Practice: Ask the classmate to your left if he or she has eaten breakfast. Ask the classmate to your right if he or she has eaten lunch.

情景(一) Scenario 1

妈妈： 早安。

Māma: Zǎo'ān.

李小明： 早上好。

Lǐ Xiǎomíng: Zǎoshang hǎo.

妈妈： 来吃早餐。

Māma: Lái chī zǎocān.

李小明： 今天早餐有什么？

Lǐ Xiǎomíng: Jīntiān zǎocān yǒu shénme?

妈妈： 今天早餐有蛋、吐司、香肠和牛奶。

Māma: Jīntiān zǎocān yǒu dàn, tǔsī, xiāngcháng, hé niúnǎi.

李小明： 太棒了！谢谢妈妈！

Lǐ Xiǎomíng: Tài bàng le! Xièxie māma!

Comprehension Questions

1. 今天早餐有什么？
2. 李小明喜欢他的早餐吗？

情景(二) Scenario 2

张爱华： 午安。

Zhāng Àihuá: Wǔ'ān.

吴家玲：下午好。你好吗？

Wú Jiālíng: Xiàwǔ hǎo. Nǐ hǎo ma?

张爱华：我很好，但是我很饿。我想吃午餐。

Zhāng Àihuá: Wǒ hěn hǎo, dànshì wǒ hěn è. Wǒ xiǎng chī wǔcān.

吴家玲：我也有点儿饿，但是我现在得去上课，没有时间吃午餐。

Wú Jiālíng: Wǒ yě yǒu diǎnr è, dànshì wǒ xiànzài děi qù shàng kè, méiyǒu shíjiān chī wǔcān.

再见！

Zàijiàn!

张爱华：再见！

Zhāng Àihuá: Zàijiàn!

Comprehension Questions

1. 张爱华想做什么？
2. 吴家玲得做什么？

情景（三）Scenario 3

奶奶去睡觉前…… Before grandmother goes to sleep...

奶奶：晚上好，爱华。晚餐吃了吗？

Nǎinai: Wǎnshang hǎo, Àihuá. Wǎncān chī le ma?

张爱华：还没吃。

Zhāng Àihuá: Hái méi chī.

奶奶：别忘了吃晚餐。晚安。

Nǎinai: Bié wàng le chī wǎncān. Wǎn'ān.

张爱华：晚安。

Zhāng Àihuá: Wǎn'ān.

Comprehension Questions

1. 张爱华晚餐吃了没?
2. 奶奶说了什么?

口语沟通活动 Oral Communication Activities

I. 情况一：午餐时间到了。林强很饿，想吃东西。

Situation 1: It's lunch time. Lin Qiang is hungry and wants to have something to eat.

任务一：两人一组。一个人扮演林强，另一个人扮演林强的妈妈。林强问他妈妈午餐有什么。林强的妈妈用下表告诉林强午餐有什么。用下面列出的会话结构帮助你们对话。

Task 1: Work in pairs. One acts as Lin Qiang and the other acts as Lin Qiang's mother. Lin Qiang asks his mother what there is for lunch. Lin Qiang's mother can use the lunch list below to tell Lin Qiang what he will have for lunch. Use the conversation structure listed below to help with your conversation.

午餐
三明治
意大利面
汉堡
牛奶
冰茶

会话结构：

Conversation Structure:

林强：妈妈，我很饿。我想吃午餐。

妈妈：来吃午餐。

林强：午餐有什么？

妈妈：午餐有＿＿＿＿＿＿＿＿＿＿＿＿＿＿。

情况二：晚餐时间到了。林强又饿了，想吃东西。

Situation 2: It's dinner time. Lin Qiang is hungry again and wants to have something to eat.

任务二：跟同一个伙伴一组。这次换角色。扮演林强的人现在是林强的妈妈。扮演林强妈妈的人现在是林强。林强问他妈妈晚餐有什么。林强的妈妈用下表告诉林强晚餐有什么。用下面列出的会话结构帮助你们对话。

Task 2: Work with the same partner, but switch roles. The one who acted as Lin Qiang is the mother and vice versa. Lin Qiang asks his mother what there is for dinner. Lin Qiang's mother can use the dinner list below to tell Lin Qiang what he will have for dinner. Use the conversation structure listed below to help with your conversation.

晚餐
炒饭
鸡肉
青菜
蛋花汤
冰茶

会话结构:

Conversation Structure:

林强：妈妈，我很饿。我想吃晚餐。

妈妈：来吃晚餐。

林强：晚餐有什么？

妈妈：晚餐有＿＿＿＿＿＿＿＿＿＿＿＿＿＿。

II. 情况：下图显示了王今的妈妈在不同时间叫王今吃的饭。

Situation: The pictures below show Wang Jin's mother telling Wang Jin to eat at different times of a day.

任务：两人一组。一个人扮演王今，另一个人扮演王今的妈妈。用下面列出的会话结构表演图片里的情景。

Task: Work in pairs. One acts as Wang Jin and the other acts as Wang Jin's mother. Use the conversation structure below to act out the scenes in the pictures.

会话结构：

Conversation Structure:

妈妈：早安。

王今：早安。

妈妈：来吃早餐。

图一 早上

图二 中午

图三 晚上

读写沟通活动 Literacy Communication Activities

I. 你的朋友林丽和你在*QQ*上聊天。她问你吃午餐了吗，以及想不想跟她一起吃午餐。你的回复已用英文写好了。把你的回复翻译成中文。

Your friend, Lin Li, and you are having a conversation on QQ. She asks if you have eaten lunch and if you want to eat lunch with her. Your replies are already written in English. Translate the replies into Chinese.

You

交谈中请勿轻信汇款、中奖信息、陌生电话，勿使用外挂软件。

林丽：你午餐吃了吗？

(Have you eaten lunch yet?)

You: ______________。

(I have not.)

You: 我们去吃午餐！

(Let's go eat lunch!)

You: ______________________________。

(I have to go to a class now, and don't have time to eat lunch.)

消息记录

关闭(C) 发送(S)

II. 你的朋友们林友、林强和林丽不同时间在脸书上跟你打招呼。请在下面回复他们。

Your friends Lin You, Lin Qiang, and Lin Li greeted you on your Facebook page at different times of day. Reply to their messages appropriately in the blanks below.

早上好！

林友　8:05 am

__________！

你的姓名：

____

午安！

林强　2:00 pm

__________！

你的姓名：

____

晚安！

林丽　9 :30 pm

__________！

你的姓名：

____

讨论 Discussion

Chinese Breakfast

What do Chinese people eat for breakfast? People from different parts of China have their own traditional breakfast dishes. However, the following items are fairly common for breakfast in China:

面条儿(miàntiáor)

粥(zhōu)

包子(bāozi)

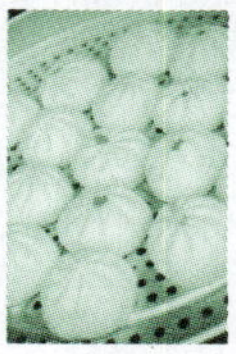

馒头(mántou)

饺子(jiǎozi)

油条(yóutiáo)

豆浆(dòujiāng)

豆腐脑(dòufunǎo)

Although there are different kinds of traditional Chinese breakfasts one can choose to eat, due to western influence, some Chinese people choose to have western breakfasts such as cereal and milk.

Discussion Questions:

1. Have you seen or tried some of the traditional Chinese breakfasts shown in the pictures above? Do you like them?
2. What other Chinese breakfasts have you seen or tried? Did you like the taste?
3. Ask your teacher or a Chinese native speaker what their favorite Chinese breakfast is.

繁体情景对话课文
Lesson Scenarios in Traditional Chinese Characters

Lesson 1

Practice: Initials

1. Lip Sounds: m p f b
2. Tip of Tongue: t d l n
3. Back of Tongue: g h k
4. Blade of Tongue: q x j
5. Curled Back of Tongue: zh r ch sh
6. Hiss Sounds: s c z

Practice: Finals

1. Simple Finals: u e i o ê a ü
2. Compound Finals
   - 2.1 ai ou ei ao
   - 2.2 ie iao iu ia
   - 2.3 ui uo uai üe ua

3. Nasal Finals and Retroflex Final

3.1 en eng ong an ang

3.2 in ian iong ing iang

3.3 uan uang ueng un(uen)

3.4 ün(üen) er üan

Practice: Spelling Rules

1. miu 2. jiu 3. xiu 4. hui 5. cun 6. sun 7. ju 8. quan 9. xun
10. ri 11. zi 12. chi 13. shi 14. ye 15. you 16. yang 17. ying
18. yong 19. wa 20. wo 21. wai 22. wei 23. wan 24. yu
25. yue 26. yuan 27. qiu 28. kun 29. gui 30. zhi

Practice: Four Tones

1.(1) bā bá bǎ bà
(2) mā má mǎ mà
(3) fā fá fǎ fà

2.(1) dā dà (2) tú tǔ (3) né nè (4) lā lǎ
(5) gē gé (6) jī jǐ (7) qí qì

The Neutral Tone

1. bàba 爸爸 (father)
2. māma 媽媽 (mother)
3. gēge 哥哥 (older brother)
4. jiějie 姐姐 (older sister)
5. dìdi 弟弟 (younger brother)
6. mèimei 妹妹 (younger sister)
7. hǎo ba 好吧 (O.K.)
8. hǎo le 好了(ready)
9. chēzi 車子 (car)

Practice: Tone Change Rules

(1) nǐ hǎo (2) lǎoshǔ (3) xiě zì (4) zǒu lù (5) bù shì
(6) bù zuò (7) yī diǎn (8) yī quān (9) yīxià (10) yī cì

Practice Chinese Pronunciations with Chinese Nursery Rhymes

1 小皮球 (Small Ball)

Xiǎo píqiú, xiāngjiāo yóu,	小皮球，香蕉油，
(Small ball, banana oil)	
Mǎn dì kāi huā èr shí yī.	滿地開花二十一。
(Flowers bloom on the ground, 21)	
èr wǔ liù, èr wǔ qī,	二五六，二五七，
(256, 257)	
èr bā, èr jiǔ, sān shí yī.	二八、二九、三十一。
(28, 29, 31)	

2 三轮车 (Tricycle Runs Fast)

Sānlúnchē, pǎo de kuài,	三輪車，跑得快，
(Tricycle runs fast)	
Hòumiàn zuò ge lǎo tàitai.	後面坐個老太太。
(An elderly woman is in the back seat)	
Yào wǔ máo, gěi yī kuài,	要五毛，給一塊，
(Asked for 50 cents, she gives 1 dollar)	
Nǐ shuō qíguài bù qíguài?	你說奇怪不奇怪？
(Don't you agree it is strange?)	

Lesson 3

Scenario 1

在中文班

李小明： 陳老師，您好！我叫李小明。我是新學生。
陳老師： 李小明，你好！
學生們： 李小明，你好！
李小明： 大家好！

Scenario 2

中文課下課後……

李小明：你好。我叫李小明。請問你叫什麼名字？
張愛華：你好，李小明。我叫張愛華。
李小明：你是哪國人？
張愛華：我是美國人。
李小明：你是美國哪裡人？
張愛華：我是美國加州人。
李小明：我是美國德州人。
張愛華：很高興認識你。
李小明：我也很高興認識你。

Lesson 4

Scenario 1

李小明：愛華，你家有幾個人？
張愛華：我家有三個人。你呢？
李小明：我家有六個人。
張愛華：你家有誰？
李小明：我家有爸爸、媽媽、兩個姐姐、一個弟弟和我。你呢？
張愛華：我家有爸爸、媽媽和我。

Scenario 2

王文麗：家玲，你有兄弟姐妹嗎？
吴家玲：有，我有一個哥哥、一個姐姐、一個弟弟和一個妹妹。你呢？
王文麗：没有，我没有兄弟姐妹，但是我有一隻狗。我家有爸爸、媽媽、我和我的狗。我愛我的家！

Lesson 5

Scenario 1

李小明： 你喜歡學中文嗎？

張愛華： 是啊！我很喜歡學中文。我每天寫漢字、讀中文書。你呢？

李小明： 我也喜歡學中文。我每天説中文。

張愛華： 我們一起練習中文好嗎？

李小明： 好啊！

Scenario 2

王文麗： 為什麼你喜歡學中文？

吴家玲： 因為我有很多中國朋友。他們都説中文。你呢？

王文麗： 因為我爸爸媽媽是中國人。

Lesson 6

Scenario 1

李小明： 愛華，明天你想做什麼？

張愛華： 我不知道。你呢？

李小明： 明天我想看電影、吃中國菜和打籃球。你想不想一起去？

張愛華： 我想看電影和吃中國菜，但是我不想打籃球。

李小明： 没關係。我們明天不打籃球。我們聽音樂怎麼樣？

張愛華： 行。我想聽音樂。

李小明： 太好了！明天見。

張愛華： 明天見。

Scenario 2

王文麗： 明天我想游泳。你想不想一起去？

吴家玲： 明天我没有空兒。今天怎麼樣？

王文麗： 今天我没有空兒。後天呢？

吴家玲: 行！後天我有空兒。我們後天去吧。

王文麗: 好,再見。

吴家玲: 再見。

Lesson 7

Scenario 1

王老師: 李小明,長大以後你想做什麼?

李小明: 長大以後我想做醫生。

王老師: 張愛華,長大以後你想做什麼?

張愛華: 我想做中文老師。

王老師: 太棒了！祝你們夢想成真!

Scenario 2

王文麗: 長大以後我想做畫家。

吴家玲: 為什麼你想做畫家?

王文麗: 因為我很喜歡畫畫兒。你呢？長大以後你想做什麼?

吴家玲: 我想做作家,因為我很喜歡寫作。

Lesson 8

Scenario 1

李小明邀請張愛華、王文麗和吴家玲到他家開派對。

李小明: 歡迎來我家！你們想喝什麼?

王文麗: 我想喝可樂。

張愛華: 我想喝水。

吴家玲: 我想喝冰茶。

李小明拿飲料給他的朋友們喝。

王文麗、張愛華和吴家玲: 謝謝!

Scenario 2

吴家玲的爺爺奶奶來吴家玲的家看她。

吴家玲: 爺爺奶奶,歡迎來我家! 請問您想喝什麼?

爺　爺: 你有什麼飲料?

吴家玲: 我有咖啡、紅茶、可樂和果汁。

爺　爺: 我想喝咖啡。

奶　奶: 我想喝紅茶。

Lesson 9

Scenario 1

李小明剛從學校回到家。

媽　媽: 今天天氣怎麼樣?

李小明: 今天早上天氣很好,但是下午有點兒冷。

媽　媽: 天氣預報説今天晚上會下雪。

李小明: 太棒了! 我可以做雪人。

Scenario 2

爺爺剛看完天氣預報。

吴家玲: 天氣預報説明天天氣怎麼樣?

爺　爺: 天氣預報説明天上午會下雨、下午會出太陽。

吴家玲: 所以明天上午是下雨天、下午是晴天。

爺　爺: 是啊,明天上午别忘了帶傘。

Lesson 10

Scenario 1

媽　媽: 早安。

李小明: 早上好。

媽　媽: 來吃早餐。

李小明: 今天早餐有什麼?

媽　媽: 今天早餐有蛋、吐司、香腸和牛奶。

李小明: 太棒了!謝謝媽媽!

Scenario 2

張愛華: 午安。

吳家玲: 下午好。你好嗎?

張愛華: 我很好,但是我很餓。我想吃午餐。

吳家玲: 我也有點兒餓,但是我現在得去上課,没有時間吃午餐。再見!

張愛華: 再見!

Scenario 3

奶奶去睡覺前……

奶　奶: 晚上好,愛華。晚餐吃了嗎?

張愛華: 還没吃。

奶　奶: 别忘了吃晚餐。晚安。

張愛華: 晚安。

词汇表
*Vocabulary Glossary

A					
啊	啊	a	particle placed at the end of a sentence for emphasis	*part.*	5
爱	愛	ài	to love	*v.*	4
B					
吧	吧	ba	particle placed at the end of a sentence to make a suggestion	*part.*	6
爸爸	爸爸	bàba	father	*n.*	4
别忘了	別忘了	bié wàng le	do not forget		9
冰茶	冰茶	bīngchá	iced tea	*n.*	8
不	不	bù	not	*adv.*	6
C					
成真	成真	chéng zhēn	to come true		7
吃	吃	chī	to eat	*v.*	6
吃了吗	吃了嗎	chī le ma	has eaten yet		10

*词汇表中的词汇为每课的生词,不包括专有名词和补充词语。

吃中国菜	吃中國菜	chī Zhōngguócài	to eat Chinese food	*v.o.*	6
出太阳	出太陽	chū tàiyáng	the sun comes out	*v.o.*	9
			D		
打	打	dǎ	to hit, to play	*v.*	6
打篮球	打籃球	dǎ lánqiú	to play basketball	*v.o.*	6
大家	大家	dàjiā	everyone	*pron.*	3
带	帶	dài	to bring	*v.*	9
但是	但是	dànshì	but	*conj.*	4
蛋	蛋	dàn	egg	*n.*	10
的	的	de	possessive particle	*part.*	4
得	得	děi	have to	*aux.*	10
弟弟	弟弟	dìdi	younger brother	*n.*	4
电影	電影	diànyǐng	movie	*n.*	6
都	都	dōu	all	*adv.*	5
读	讀	dú	to read	*v.*	5
多	多	duō	many; a lot of	*adj.*	5
			E		
饿	餓	è	hungry	*adj.*	10
			G		
高兴	高興	gāoxìng	glad, happy	*adj.*	3
哥哥	哥哥	gēge	elder brother	*n.*	4
个	個	gè	general classifier	*m.w.*	4
狗	狗	gǒu	dog	*n.*	4

果汁	果汁	guǒzhī	juice	*n.*	8
H					
汉字	漢字	Hànzì	Chinese character	*n.*	5
好	好	hǎo	good	*adj.*	3
好	好	hǎo	great (expressing agreement)	*adj.*	5
好吗	好嗎	hǎo ma	is it all right with someone (asking for agreement)		5
喝	喝	hē	to drink	*v.*	8
和	和	hé	and	*conj.*	4
很	很	hěn	very	*adv.*	3
红茶	紅茶	hóngchá	black tea	*n.*	8
后天	後天	hòutiān	day after tomorrow	*n.*	6
画画儿	畫畫兒	huà huàr	to paint	*v.o.*	7
画家	畫家	huàjiā	painter	*n.*	7
还	還	hái	still	*adv.*	10
欢迎	歡迎	huānyíng	to welcome	*v.*	8
会	會	huì	will	*aux.*	9
J					
几	幾	jǐ	how many	*num.*	4
家	家	jiā	home, family	*n.*	4
见	見	jiàn	to see, to meet	*v.*	6
叫	叫	jiào	to call	*v.*	3
姐姐	姐姐	jiějie	older sister	*n.*	4
今天	今天	jīntiān	today	*n.*	6

K					
咖啡	咖啡	kāfēi	coffee	*n.*	8
看	看	kàn	to see, to look	*v.*	6
看电影	看電影	kàn diànyǐng	to see a movie	*v.o.*	6
可乐	可樂	kělè	coke	*n.*	8
可以	可以	kěyǐ	can	*aux.*	9
空儿	空兒	kòngr	free time	*n.*	6
L					
来	來	lái	to come	*v.*	8
篮球	籃球	lánqiú	basketball	*n.*	6
老师	老師	lǎoshī	teacher	*n.*	3
冷	冷	lěng	cold, chilly	*adj.*	9
练习	練習	liànxí	to practice	*v.*	5
了	了	le	particle used to indicate completion of an action	*part.*	10
M					
妈妈	媽媽	māma	mother	*n.*	4
吗	嗎	ma	particle used to form a yes or no question	*part.*	4
没关系	没關係	méi guānxi	never mind		6
没有	没有	méiyǒu	to not have	*v.*	4
每	每	měi	every	*pron.*	5
妹妹	妹妹	mèimei	younger sister	*n.*	4
们	們	men	plural for pronouns and human nouns	*suff.*	3

梦想	夢想	mèngxiǎng	dream	*n.*	7
梦想成真	夢想成真	mèngxiǎng chéng zhēn	dreams come true		7
名字	名字	míngzi	name	*n.*	3
明天	明天	míngtiān	tomorrow	*n.*	6
N					
哪	哪	nǎ	which	*pron.*	3
哪里	哪裏	nǎlǐ	where	*pron.*	3
奶奶	奶奶	nǎinai	grandmother	*n.*	8
呢	呢	ne	question particle	*part.*	4
你	你	nǐ	you	*pron.*	3
你们	你們	nǐmen	you (plural)	*pron.*	7
您	您	nín	you (polite form)	*pron.*	3
牛奶	牛奶	niúnǎi	milk	*n.*	10
P					
朋友	朋友	péngyou	friend	*n.*	5
Q					
天气预报	天氣預報	tiānqì yùbào	weather forecast		9
晴天	晴天	qíngtiān	sunny day	*n.*	9
请问	請問	qǐngwèn	May I ask... (to show politeness when asking a question)	*v.*	3
去	去	qù	to go	*v.*	6
R					
人	人	rén	people, person	*n.*	3
认识	認識	rènshi	to know, to be acquainted	*v.*	3

S					
伞	傘	sǎn	umbrella	*n.*	9
上课	上課	shàng kè	to go to class	*v.o.*	10
上午	上午	shàngwǔ	moring	*n.*	9
什么	什麼	shénme	what	*pron.*	3
时间	時間	shíjiān	time	*n.*	10
是	是	shì	to be	*v.*	3
书	書	shū	book	*n.*	5
谁	誰	shéi	who	*pron.*	4
水	水	shuǐ	water	*n.*	8
说	說	shuō	to speak	*v.*	5
所以	所以	suǒyǐ	therefore, so	*conj.*	9
T					
他	他	tā	he, him	*pron.*	3
她	她	tā	she, her	*pron.*	3
太棒了	太棒了	tài bàng le	great		7
太好了	太好了	tài hǎo le	great		6
天气	天氣	tiānqì	weather	*n.*	9
听	聽	tīng	to listen	*v.*	6
听音乐	聽音樂	tīng yīnyuè	to listen to music	*v.o.*	6
吐司	吐司	tǔsī	toast	*n.*	10
W					
晚安	晚安	wǎn'ān	good night		10

晚餐	晚餐	wǎncān	dinner	*n.*	10
晚上	晚上	wǎnshang	evening	*n.*	9
晚上好	晚上好	wǎnshang hǎo	good evening		10
为什么	為什麼	wèi shénme	why		5
我	我	wǒ	I, me	*pron.*	3
我们	我們	wǒmen	we, us	*pron.*	5
午安 下午好	午安 下午好	wǔ'ān xiàwǔ hǎo	good afternoon		10
午餐	午餐	wǔcān	lunch	*n.*	10
X					
喜欢	喜歡	xǐhuan	to like	*v.*	5
下午	下午	xiàwǔ	afternoon	*n.*	9
下雪	下雪	xià xuě	to snow	*v.o.*	9
下雨	下雨	xià yǔ	to rain	*v.o.*	9
下雨天	下雨天	xià yǔ tiān	rainy day		9
现在	現在	xiànzài	now	*n.*	10
香肠	香腸	xiāngcháng	sausage	*n.*	10
想	想	xiǎng	to want	*v.*	6
写	寫	xiě	to write	*v.*	5
写作	寫作	xiězuò	to write	*v.*	7
谢谢	謝謝	xièxie	to thank	*v.*	8
新	新	xīn	new	*adj.*	3
行	行	xíng	OK	*v.*	6
兄弟姐妹	兄弟姐妹	xiōngdì jiěmèi	siblings		4

学	學	xué	to study	*v.*	5
学生	學生	xuéshēng	student	*n.*	3
雪人	雪人	xuěrén	snowman	*n.*	9
Y					
爷爷	爺爺	yéye	grandfather	*n.*	8
也	也	yě	also, too	*adv.*	3
一起	一起	yìqǐ	together	*adv.*	5
医生	醫生	yīshēng	medical doctor	*n.*	7
以后	以後	yǐhòu	after	*n.*	7
因为	因為	yīnwèi	because	*conj.*	5
音乐	音樂	yīnyuè	music	*n.*	6
饮料	飲料	yǐnliào	drink	*n.*	8
游泳	游泳	yóu yǒng	to swim		6
有	有	yǒu	to have	*v.*	4
有点儿	有點兒	yǒudiǎnr	a bit	*adv.*	9
Z					
再见	再見	zàijiàn	goodbye	*v.*	6
早安 早上好	早安 早上好	zǎo'ān zǎoshang hǎo	good morning		10
早餐	早餐	zǎocān	breakfast	*n.*	10
怎么样	怎麼樣	zěnmeyàng	how	*pron.*	6
长大	長大	zhǎngdà	to grow up		7
知道	知道	zhīdào	to know	*v.*	6
只	隻	zhī	classifier for animals	*m.w.*	4

中国菜	中國菜	Zhōngguócài	Chinese food, Chinese dish	*n.*	6
中文	中文	Zhōngwén	Chinese language	*n.*	5
祝	祝	zhù	to wish	*v.*	7
作家	作家	zuòjiā	writer	*n.*	7
做	做	zuò	to do	*v.*	6

We Speak Chinese

我们说中文

练习册 初级 1

Workbook Beginner

宋可音（Ko-Yin Sung）编著

北京大学出版社
PEKING UNIVERSITY PRESS

目录
Contents

第一课
Lesson 1

汉语拼音
The Chinese Phonetic System—Hanyu Pinyin

I. 拼音练习 Pinyin Exercises

1. Initials

(1) Listen to the audio recordings and repeat after them

(2) Listen to the audio recordings and circle the syllables you hear

1. bù	pù	2. mó	fó	3. dí	tí
4. mǎ	nǎ	5. lì	nì	6. gē	kē
7. jú	qú	8. zhè	zè	9. shī	sī
10. chā	cā				

(3) Listen to the audio recordings and write down the initials you hear. Each of the 21 initials only appears once

1. ___ú	2. ___ǔ	3. ___è	4. ___ó	5. ___ū	6. ___à
7. ___à	8. ___í	9. ___è	10. ___ā	11. ___í	12. ___ù
13. ___ì	14. ___ǐ	15. ___è	16. ___ǐ	17. ___ū	18. ___ú
19. ___ī	20. ___ǐ	21. ___ē			

2. Finals

(1) Listen to the audio recordings and repeat after them

(2) Listen to the audio recordings and circle the syllables you hear

1. nǔ	nǚ	2. duō	dōu	3. duì	diù
4. zǎo	zhǎo	5. xióng	xiáng	6. zàn	zàng
7. hěn	hēng	8. nín	níng	9. dùn	duàn
10. jiā	jiāo				

(3) Listen to the audio recordings and write down the finals you hear. Each of the finals only appears once

1. d___	2. b___	3. s___	4. l___	5. d___
6. n___	7. l___	8. d___	9. p___	10. t___
11. x___	12. x___	13. b___	14. l___	15. h___
16. g___	17. g___	18. h___	19. l___	20. s___
21. b___	22. t___	23. d___	24. z___	25. t___
26. m___	27. n___	28. t___	29. q___	30. t___
31. k___	32. sh___	33. w___	34. x___	35. j___
36. zh___				

II. Pinyin Spelling and Tone Mark Exercises

1. Tone Marks

Listen to the audio recordings and mark the tones you hear

1. zan	2. tang	3. xiong	4. ji	5. shuang
6. zhao	7. jie	8. mei	9. dou	10. jiu
11. chuan	12. qi			

2. Pinyin Spelling with Tone Marks

Listen to the audio recordings and write down the syllables and tones you hear

1. ________	2. ________	3. ________	4. ________
5. ________	6. ________	7. ________	8. ________
9. ________	10. ________	11. ________	12. ________

13. ________ 14. ________ 15. ________ 16. ________
17. ________ 18. ________ 19. ________ 20. ________
21. ________ 22. ________ 23. ________ 24. ________
25. ________ 26. ________ 27. ________ 28. ________

III. Tone Change Exercises

1. Third Tone Change Rules

Rule 1: 3rd + 3rd are pronounced as 2nd + 3rd

Rule 2: 3rd + 1st, 2nd, or 4th are pronounced as half 3rd + 1st, 2nd, or 4th

Listen to and repeat after the audio recordings to practice the third tone change rules. When you finish, read aloud the following words to your teacher or a classmate.

1. qǐng xiě 请写 (please write)
2. zǒu zǒu 走走 (take a walk)
3. xiǎng nǐ 想你 (miss you)
4. nǐ hǎo 你好 (hi)
5. nǐ zǎo 你早(good morning)
6. wǒ qù 我去 (I go)
7. hěn lèi 很累 (very tired)
8. qǐng wèn 请问 (may I ask)
9. zǒng shì 总是(always)
10. yě duì 也对 (also correct)

2. Tone Change Rule about the Pronunciation of 不 (bù)

Rule: 不(bù) + 4th are pronounced as 不(bú) + 4th

Listen to and repeat after the audio recordings to practice the tone change rule about the pronunciation of 不 (bù). When you finish, read aloud the following words to your teacher or a classmate for more practice.

1. bú shì 不是 (is not)
2. bú ài 不爱 (not love)
3. bú duì 不对 (not correct)
4. bú zuò 不做 (not do)
5. bú dìng 不定 (not certain)
6. bú kàn 不看 (not look)
7. bú dà 不大 (not big)
8. bú yào 不要 (not want)
9. bú lì 不利 (unfavorable)
10. bú zài 不再 (not anymore)

3. Tone Change Rules about the Pronunciations of 一 (yī)

Rule 1: 一 (yī) + 1st, 2nd, or 3rd are pronounced as 一 (yì) + 1st, 2nd, or 3rd

Rule 2: 一 (yī) + 4th are pronounced as 一 (yí) + 4th

Listen to and repeat after the audio recordings to practice the tone change rules about the pronunciation of 一 (yī). When you finish, read aloud the following words to your teacher or a classmate.

1. yìbiān 一边 (a side)
2. yìtiān 一天 (a day)
3. yìqiān 一千 (a thousand)
4. yì píng 一瓶 (a bottle)
5. yìdiǎn 一点 (a dot)
6. yí bàn 一半 (a half)
7. yí zì 一字 (a word)
8. yíxià 一下 (a brief moment)
9. yí biàn 一遍 (once)
10. yí cì 一次 (once)

第二课
Lesson 2

汉字系统
The Chinese Writing System

I. Strokes Exercise

Write the eight basic strokes in the blank boxes provided below

1. 横 héng Horizontal Stroke	2. 竖 shù Vertical Stroke	3. 撇 piě Left Leg	4. 捺 nà Right Leg
5. 提 tí Rising Stroke	6. 点 diǎn Dot Stroke	7. 钩 gōu Hook Stroke	8. 折 zhé Bend Stroke

II. Stroke Order Exercise

Use the hints provided to you below to write the following characters in the right stroke order. The first character is done for you as an example

1. 十

Hint: First horizontal, then vertical

First stroke: 一

Second stroke: 十

2. 古

Hint: Top to bottom

First stroke:

Second stroke:

Third stroke:

Fourth stroke:

Fifth stroke:

3. 什

Hint: Left to right

First stroke:

Second stroke:

Third stroke:

Fourth stroke:

4. 田

Hint: Outside first, then inside, then close the bottom side last

First stroke:

Second stroke:

Third stroke:

Fourth stroke:

Fifth stroke:

5. 大

Hint: Left leg first, then right leg

First stroke:

Second stroke:

Third stroke:

6. 小

Hint: Middle first, then the sides

First stroke:

Second stroke:

Third stroke:

III. Character Structures

1. Radicals

(1) The radicals listed below can be used as independent characters. Write down the meanings in English below the characters

车　　口　　心　　马　　人

______　______　______　______　______

女　　火　　雨　　门　　木

______　______　______　______　______

(2) The 25 characters in the table below are numbered from 1 to 25. Identify the radical of each character and write down their numbers next to the radicals they have. The characters which have “车” as the radical have been identified for you as an example

1	吗	6	闪	11	烧	16	忐	21	闭
2	雪	7	人	12	忑	17	林	22	吧
3	姑	8	驱	13	呢	18	女	23	转
4	怎	9	娜	14	门	19	你	24	妹
5	杖	10	煤	15	骑	20	轨	25	树

车：20, 23 口：______ 心：______
马：______ 人：______ 女：______
火：______ 雨：______ 门：______
木：______

2. Compound Character Structures

The compound characters listed below have one of the two-part or three-part structures introduced in Lesson 2 in Beginner 1 Textbook. Identify the structure of each character and draw their structures in the blanks above the characters. The structure of the first character “谢” is completed for you as an example.

Character Structures				
	谢	他	忘	逃
Character Structures				
	回	草	娱	筋
Character Structures				
	炒	过	国	态

第三课 Lesson 3

很高兴认识你！It's a Pleasure to Make Your Acquaintance!

I. 听力练习 Listening Exercise

今天是王老师上中文课的第一天。听他和学生的对话并和一位伙伴讨论下列问题。

Today is the first day Teacher Wang teaches Chinese class. Listen to his conversation with his students and discuss the questions below with a partner.

问题 Questions:

1. 王老师是哪国人？
   Where is teacher Wang from?
2. 陈国是美国哪里人？
   Where in the United States is Chen Guo from?
3. 林丽是美国哪里人？
   Where in the United States is Lin Li from?

II. 会话练习 Speaking Exercise

下面三个图片列出来自不同国家的人。两人一组，轮流互相问答，“他是哪国人？”把对的答案圈起来。

The three pictures below show people from different countries. Work in pairs. Take turns to ask each other the question “他是哪国人?”. Circle the right answer below the pictures.

中国人 or 美国人

加拿大人 or 美国人

中国人 or 加拿大人

下面三个图片列出来自不同州的美国人和不同城市的中国人。两人一组，轮流互相问答：“他是美国哪里人？”“他是中国哪里人？”把对的答案圈起来。

The three pictures below show American people from different states and Chinese people from different cities. Work in pairs. Take turns to ask each other the questions “他是美国哪里人？”and “他是中国哪里人？” Circle the right answer below the picture.

美国加州人 or 美国德州人

中国北京人 or 中国南京人

中国北京人 or 中国上海人

III. 阅读练习 Reading Exercise

你在你的生日派对遇到了五个新朋友。他们是王小新、张大新、李小明、张老师和李中。他们每个人都是从不同地方来的人。根据下面的图片回答他们是哪里人等问题。

You met 5 new friends at your birthday party. They are Wang Xiaoxin, Zhang Daxin, Li Xiaoming, Zhang Laoshi, and Li Zhong. Each of them is from a different place in the world. Use the pictures shown below to remind you where they are from and answer the questions below the pictures.

1. 王小新

2. 张大新

3. 李小明

4. 张老师

5. 李中

问题 Questions:

1. 谁是德州人？
   Who is from Texas?
2. 张老师是中国人吗？
   Is Teacher Zhang Chinese?
3. 王小新是加拿大人吗？
   Is Wang Xiaoxin from Canada?
4. 李中是哪里人？
   Where is Li Zhong from?
5. 张大新是加拿大哪里人？
   Where in Canada is Zhang Daxin from?

IV. 写作练习 Writing Exercise

为了要认识你，你的新中国朋友林文信里写了两个问题。请回林文的信。先跟他打招呼，然后回答他的两个问题，再用中文写"很高兴认识你"。最后在信尾写上你的名字。

In order to get to know you, your new friend from China, Lin Wen, wrote two questions for you in his letter (see the questions below). Reply to Lin Wen by saying hi first, and then answer his two questions, and write "It's a pleasure to make your acquaintance!" with your name at the end of the letter.

林文的问题　Lin Wen's questions for you:

1. 请问你叫什么名字？
2. 你是美国哪里人？

Your Reply

林文：

__________！(Hi)

我叫 ________。我是______________________人。

(your name) (country + state or city)

很高兴____________________你！

(to know)

____

(your name)

V. 沟通练习 Communicative Exercise

带一张你喜欢的人的图片到班上。给大家看这张图片并告诉大家这个人的名字和这个人是哪里人。把这个人的国家名用中文写在黑板上。

Bring a picture of a person you like to class. When you show the picture to the class, tell the class the name of the person and where the person is from. Write the name of the person's country in Chinese on the board.

第四课
Lesson 4

我爱我的家
I Love My Family

I. 听力练习 Listening Exercise

王红是中文班的新学生。她正在跟陈国谈她的家人。听他们的对话并和一位伙伴讨论下列问题。

Wang Hong is a new student in Chinese class. She is talking about her family with Chen Guo. Listen to their conversation and discuss the questions below with a partner.

问题 Questions:

1. 王红家有几个人？王红家有谁？
   How many people are in Wang Hong's family? Who are they?
2. 陈国家有几个人？陈国家有谁？
   How many people are in Chen Guo's family? Who are they?
3. 谁的家有狗？有几只狗？
   Who has dogs at home? How many dogs does that person have?
4. 谁的家有猫？
   Who has a cat at home?

II. 会话练习 Speaking Exercise

画出你未来的家。你的家会有几个人？他们是谁？你的家会有宠物吗？当你准备好了，给一位伙伴看你的图并描述你未来的家庭。

Draw your future family. How many family members will you have? Who are they?

Will there be any pets in your future family? After you finish drawing, show a partner your future family and describe how many members will be in your future family and who they are.

把你未来的家画在下面：

Draw your future family in the box below:

III. 阅读练习 Reading Exercise

王中、林文、陈丽和李文为了他们的中文作业画了家庭树。他们的最后一个任务是写出他们的家庭成员。他们每个人都忘了写家里的某些成员。你的任务是检查他们的家庭树、读他们写的答案并指出他们忘了哪些家庭成员。

Wang Zhong, Lin Wen, Chen Li, and Li Wen have made a family tree for their Chinese homework. Their last task in the homework is to write down their family members. All of them missed one or more of their family members in the answers. Your task is to check their family trees, read their answers and identify which family members are missing in the answers.

王中的家

王中：我家有爸爸、妈妈和我。

王中忘了哪个家庭成员？ ________________

Which family member is missing in Wang Zhong's sentence above?

____

林文的家

林文：我家有妈妈、一个妹妹和我。
林文忘了哪个家庭成员？________
Which family member is missing in Lin Wen's sentence above?
____

陈丽的家

陈丽：我家有一个姐姐、两个弟弟和我。
陈丽忘了哪些家庭成员？________
Which family members are missing in Chen Li's sentence above?
____

李文的家

李文：我家有爸爸、妈妈、两个哥哥和我。
李文忘了哪个家庭成员？________

Which family member is missing in Lin Wen's sentence above?

____

IV. 写作练习 Writing Exercise

你收到了你的中国朋友林文的信。信里问你家有几个人等。回林文的信并回答他的问题。

You received a letter from your Chinese friend, Lin Wen, in which he asked how many family members you have and who they are. Reply to Lin Wen and answer his questions about your family.

Your Reply

林文：

你好！

我家有 ________ 个人。我家有 ________________ 和我。
(family members)

我爱我的家！

____
(your name)

V. 沟通练习 Communicative Exercise

有三个人搬到了一个新社区。他们在社区里散步并做自我介绍。想象你是这三个人中的一个，你的教室是新社区，你的同学是你的新邻居。自己选择扮演下列三人之一。在教室里走动并用中文向其他人做自我介绍。做完自我介绍，请你的同学猜一猜你是三个人中的哪一个。

Three people moved to a new neighborhood. They walk around the neighborhood to introduce themselves to their neighbors. Imagine that you are one of the three newcomers, that your classroom is the new neighborhood, and that your classmates are your new neighbors. Pick one of the profiles below and act as him or her. Walk around the classroom to introduce yourself in Chinese using the profile. After you introduce yourself, ask your classmates who talk to you to guess which profile you chose.

1. 小王

I am from Beijing, China.
I have a dog.
I have three younger sisters.
I love my family.

2. 小李

I am from California, in the United States.
I have a cat.
I have no siblings.
I have a father and a mother.

3. 小陈

I am Chinese.
I have two elder brothers.
I have two cats and two dogs.
I love my family.

第五课 Lesson 5

我喜欢学中文
I Like Studying Chinese

I. 听力练习 Listening Exercise

李丽和陈国正在谈论英文学习。听他们的对话并和一位伙伴讨论下列问题。

Li Li and Chen Guo are talking about learning English. Listen to their conversation and discuss the questions below with a partner.

问题 Questions:

1. 为什么李丽喜欢学英文？
   Why does Li Li like to study English?
2. 为什么陈国喜欢学英文？
   Why does Chen Guo like to study English?
3. 陈国每天怎么练习英文？
   What does Chen Guo do to practice English every day?
4. 李丽在对话结束前对陈国做了什么提议？
   What is Li Li's suggestion to Chen Guo at the end of their conversation?

II. 会话练习 Speaking Exercise

吴丽的家人会说多国语言。每天她的家人喜欢读不同语言的书。两人一组。看下面的图片并告诉你的伙伴每天每个家庭成员读什么书。

Wu Li’s family is multi-lingual. Every day the family members like to read books in different languages they know. Work in pairs. Look at the pictures below and tell your partner what each family member is reading every day.

爸爸 	妈妈 
哥哥 	吴丽 

III. 阅读练习 Reading Exercise

你收到了你的中国朋友林文的信。他在信中谈到他对学法文的兴趣。读林文的信并回答下面的问题。

You received a letter from your Chinese friend, Lin Wen, in which he mentioned his interest in learning French. Read Lin Wen's letter and answer the questions below.

> 你好!
>
> 因为我爸爸是法国人,我很喜欢学法文。我每天说法文、读法文书。你呢?
>
> 你喜欢学法文吗?我们一起练习法文好吗?
>
> 林文

问题 Questions:

1. 为什么林文喜欢学法文?
   Why does Lin Wen like to study French?
2. 林文每天怎么练习法文?
   What does Lin Wen do to practice French every day?
3. 林文在信里问了你什么有关学外语的问题?
   What are the questions about learning a foreign language Lin Wen asked you in his letter?

IV. 写作练习 Writing Exercise

回林文的信。先跟林文打招呼,再回答他有关学外语的问题。

Reply to Lin Wen by saying hi first, and then answer his questions to you about learning a foreign language.

Your Reply

林文：

____________！

____

____

____

____

(your name)

V. 沟通练习 Communicative Exercise

听中文音乐、写汉字、说中文和读中文书是汉语学生学汉语常做的活动。在教室走动并问十个同学他们喜不喜欢上述的活动。用以下的句型："你喜欢_______吗？"如果你的同学回答："是啊！我很喜欢。"在他的名字旁和活动栏里打钩。如果你的同学回答："我不喜欢。"打叉。

Listening to Chinese music, writing Chinese characters, speaking Chinese and reading Chinese books are activities Chinese language students usually do to improve their Chinese. Walk around the classroom and ask ten of your classmates if they like each of the four activities using the sentence structure "你喜欢____________吗？". If your classmates answer, "是啊！我很喜欢。", put a check mark next to their name and under the activity name in the survey below. If your classmates answer, "我不喜欢。", put an "×" instead.

名字	听中文音乐	写汉字	说中文	读中文书
1.				
2.				
3.				
4.				
5.				
6.				
7.				
8.				
9.				
10.				

调查后的问题：

After-Survey Questions:

有几个人喜欢上述活动？

How many people like each of the four activities?

1. 几个人喜欢听中文音乐？__________

2. 几个人喜欢写汉字？__________

3. 几个人喜欢说中文？__________

4. 几个人喜欢读中文书？__________

第六课
Lesson 6

明天你想做什么？
What Do You Want to Do Tomorrow?

I. 听力练习 Listening Exercise

李丽正在告诉陈国她明天的计划。听他们的对话并和一位伙伴讨论下列问题。

Li Li is telling Chen Guo about her plans for tomorrow. Listen to their conversation and discuss the questions below with a partner.

问题 Questions:

1. 李丽明天要做什么？
   What is Li Li going to do tomorrow?
2. 陈国觉得他的中文好吗？
   Does Chen Guo think his Chinese is good?
3. 李丽的中国朋友会说什么语言？
   What language(s) does Li Li's Chinese friend speak?
4. 陈国接受李丽的邀请了吗？
   Did Chen Guo accept Li Li's invitation?

II. 会话练习 Speaking Exercise

把你星期六和星期日的活动计划画出来。当你准备好了，找一个伙伴并告诉他你的计划。问你的伙伴有没有时间，还有他想不想参加你的活动。如果你的伙伴拒绝你，问其他的同学，直到有人接受你的邀请。

Draw illustrations of your plans on Saturday and Sunday. When you finish, find a partner and tell them what you want to do Saturday and Sunday. Ask your partner if he or she has free time and if he or she wants to do the activities with you. If your partner declines your invitation, talk to others in class until someone accepts your invitation.

星期六活动计划 Saturday Plans	星期日活动计划 Sunday Plans

III. 阅读练习 Reading Exercise

李丽写了有关陈国和她明后天计划的讨论。读李丽的日记并回答下列问题。

Li Li wrote about her discussion with Chen Guo about hanging out together tomorrow and the day after tomorrow. Read Li Li's diary and answer the questions below.

陈国和我是好朋友。明天陈国想和我一起去游泳。我明天有空儿，但是我不想游泳。后天我想和陈国一起打篮球。陈国说后天他想看电影，但是不想打篮球。

问题 Questions:

1. 陈国明天想和李丽做什么？
   What does Chen Guo want to do with Li Li tomorrow?
2. 李丽明天有时间吗？
   Does Li Li have free time tomorrow?
3. 李丽后天想和陈国做什么？
   What does Li Li want to do with Chen Guo the day after tomorrow?
4. 陈国后天想做什么？
   What does Chen Guo want to do the day after tomorrow?

IV. 写作练习 Writing Exercise

写一篇短邀请函。在邀请函里提一个活动计划并邀请你的朋友一起来参加。

Write a short invitation note to a friend. In the invitation ask your friend if he or she wants to do activities together.

____

____

____

____

____

____

V. 沟通练习 Communicative Exercise

小组练习。组员轮流当领导。领导得问出每个组员明天想做什么并提一个每个组员喜欢的活动计划。记得用以下问句:“明天你想做什么?”“明天你有没有空儿?”“你想不想一起去?”

Work in small groups. Take turns to be the activity organizer. As an activity organizer, you need to find out what activities each of your group members feel like doing tomorrow and propose an activity plan everyone in the group agrees to do. Be sure to ask questions such as “明天你想做什么?” “明天你有没有空?” and “你想不想一起去?”

第七课
Lesson 7

长大以后你想做什么？
What Do You Want to Be When You Grow Up?

I. 听力练习 Listening Exercise

李老师正在和张爱林讨论她梦想中的工作。听他们的对话并和一位伙伴讨论下列问题。

Teacher Li is asking Zhang Ailin about her dream job. Listen to their conversation and discuss the questions below with a partner.

问题 Questions:

1. 张爱林长大以后想做什么？
   What does Zhang Ailin want to do when she grows up?
2. 张爱林为什么想做她说的梦想工作？
   What are the reasons Zhang Ailin wants to do the job she mentioned in the conversation?
3. 在对话的最后，李老师对张爱林说了什么？
   What did Teacher Li say to Zhang Ailin at the end of the conversation?

II. 会话练习 Speaking Exercise

李师有五个孩子：李林、李爱、李丽、李华和李明。每个孩子都有他们长大后想做的梦想工作。下面图片显示他们每个人的梦想工作。小组练习。一个人扮演李师，其他人扮演孩子。李师一个一个问他的孩子长大以后想做什么。孩子们根据下面图片回答他们的爸爸。

Li Shi has five children, Li Lin, Li Ai, Li Li, Li Hua, and Li Ming, all of whom have dream jobs they want to do when they grow up. The dream jobs the children like are shown in the pictures below. Work in small groups. One acts as Li Shi and the rest act as the children. Li Shi will ask his children one by one what they want to do when they grow up. The children can use the pictures below to answer their father.

李林

李爱

李丽

李华

李明

III. 阅读练习 Reading Exercise

你的中国朋友林文写了有关他和他弟弟长大以后想做的工作并问你以后的梦想工作是什么。读林文的信并回答下列问题。

Your Chinese pen pal, Lin Wen, wrote to you about what he and his younger brother would like to do when they grow up and to ask you about your dream job in the future. Read Lin Wen's letter and answer the questions below.

你好！

长大以后我想做运动员，因为我很喜欢运动。长大以后我弟弟想做老师，因为他很喜欢教书。长大以后你想做什么？

林文

问题 Questions:

1. 林文长大以后想做什么？
   What does Lin Wen want to do when he grows up?

2.林文想做他在信里提到的工作的原因是什么？
What is the reason Lin Wen wants to do the job he mentioned in the letter?
3. 林文的弟弟长大以后想做什么？
What does Lin Wen's brother want to do when he grows up?
4. 林文的弟弟想做他在信里提到的工作的原因是什么？
What is the reason Lin Wen's brother wants to do the job mentioned in the letter?

IV. 写作练习 Writing Exercise

回林文的信并回答他的问题。在信的最后祝林文梦想成真。

Reply to Lin Wen and answer his question about your dream job. At the end of your letter, express a wish for Lin Wen's dreams to come true.

Your Reply

林文：

你好！

我长大以后____________。因为我很喜欢____________。

祝你____________！

____

V. 沟通练习 Communicative Exercise

准备一小段演讲稿。说说你长大以后的梦想工作和为什么你想做这样的工作。当你准备好了，你的老师会请你在班上演讲。演讲完后，你的同学们会祝你梦想成真。

Prepare a short speech in which you talk about your dream job and why you want to do it when you grow up. When you are ready, your teacher will ask you to give the speech in front of the class. When you finish your speech, your classmates will express a wish for your dreams to come true.

第八课
Lesson 8

你想喝什么？
What Would You Like to Drink?

I. 听力练习 Listening Exercise

王国和他的父母正在拜访王国的老师林老师。听他们的对话并和一位伙伴讨论下列问题。

Wang Guo and his parents are visiting Wang Guo's teacher, Teacher Lin. Listen to their conversation and discuss the questions below with a partner.

问题 Questions:

1. 林老师有什么饮料？
   What drinks does Teacher Lin have?
2. 王国的爸爸想喝什么？
   What would Wang Guo's father like to drink?
3. 王国的妈妈想喝什么？
   What would Wang Guo's mother like to drink?
4. 王国想喝什么？
   What would Wang Guo like to drink?

II. 会话练习 Speaking Exercise

小组练习。一个人扮演主人，其他人扮演客人。主人今晚有一个派对。他得欢迎他的朋友们来他家拜访。主人得问客人想喝什么。客人得问主人有什么饮料。主人要告诉客人有什么饮料并让他们选择想喝的饮料。

Role play the following scenario in small groups. One acts as the host and the rest are visitors. The host has a party at home this evening. He or she should welcome his or her friends to his or her home (欢迎来我家). The host should then ask them what they would like to drink (你想喝什么？). The friends will ask the host what drinks he or she has (你有什么饮料？). The host will tell them the drinks available (我有__________) and let them choose what they want.

III. 阅读练习 Reading Exercise

两人一组。一起读图片里的对话并回答下面的问题。

Work in pairs. Read the dialogues in the pictures together and discuss the answers to the questions below.

Question: 妈妈想喝什么？

Question: 奶奶有可乐吗？

Question: 弟弟有什么饮料？　　Question: 陈老师想喝什么？

IV. 写作练习 Writing Exercise

你的第一间饮料店要开张了。开店前你需要设计饮料单和决定你要卖哪些饮料。两人一组。用下面的空格设计饮料单。设计完了以后给班上同学看。同学会问你："你有XXX(饮料名)？"你要告诉他们饮料单里有哪些饮料。

You are going to open your first drink shop. Before you open your shop, you need to design the drink menu and decide which beverages you are going to sell. Work in pairs and use the space provided below to create your menu. After you complete the menu, show it to the class. The class will ask you "你有什么饮料？", and you are going to tell the class which drinks are listed on the menu.

V. 沟通练习 Communicative Exercise

你在一间饮料店买饮料给你的家人。两人一组。轮流扮演客人和饮料店老板。客人得告诉老板他的每个家人要喝什么饮料。客人点完饮料后老板要重复说一次客人的订单并确认订单是正确的。

You are in a drink shop buying drinks for your family. Work in pairs. Take turns to role play the customer and shop owner. The customer will tell the owner which drink each of his or her family members would like to drink. When the customer finishes ordering, the owner needs to repeat the order to confirm that the information is correct.

第九课
Lesson 9

今天天气怎么样？
How is the Weather Today?

I. 听力练习 Listening Exercise

一个北京的中文气象播报员正在预报天气。听他的预报并和一位伙伴讨论下列问题。

A Chinese weatherman in Beijing is announcing the weather forecast. Listen to his report and discuss the questions below with a partner.

问题 Questions:

1. 北京明天会下雪吗？
   Will it snow tomorrow in Beijing?
2. 北京明天的天气怎么样？
   What will the weather be in Beijing tomorrow morning?
3. 明天下午北京会出太阳吗？
   Will it be sunny tomorrow afternoon in Beijing?
4. 明天晚上北京的天气怎么样？
   What will the weather be in Beijing tomorrow evening?

II. 会话练习 Speaking Exercise

做这个活动前，先听听当地的天气预报。把你听到的天气情况画在下面的空格中。明天和后天的天气怎么样？当你准备好了，口头描述你画的图给一个伙伴听。

Listen to the weather forecast in the local news before you do this exercise. Based on the information you hear, draw an illustration of the kinds of weather forecast for tomorrow morning, tomorrow evening, the morning after tomorrow and the evening after tomorrow in the boxes provided below. When you complete the drawing, describe what you have drawn to a partner.

明天上午	明天晚上	后天上午	后天晚上

III. 阅读练习 Reading Exercise

你错过了今天早上的天气预报，但是你妈妈把天气预报写在下面一张纸条上。两人一组。大声朗读下面的纸条并一起讨论下列问题。

You missed today's weather report in the morning while you were sleeping, but your mother has written down a note about the report. Work in pairs. Read aloud the note together and discuss the answers to the questions below the message.

> 天气预报说今天上午会下雨，下午会出太阳，晚上会下雪。
>
> 早上别忘了带伞。
>
> 妈妈

问题 Questions:

1. 今天上午天气怎么样？
   How is the weather this morning?
2. 今天下午天气怎么样？
   What will the weather be this afternoon?
3. 今天晚上天气怎么样？
   What will the weather be this evening?
4. 你妈妈在纸条里提醒了你什么？
   What did your mother remind you to do in the note?

IV. 写作练习 Writing Exercise

你是一个中文天气播报员。你要用中文预报中国四个主要城市的天气。你要预报的信息已为你在下面用英文准备好了。你得把英文翻译成中文才能做预报。把你的中文翻译写在英文下面。

You are a Chinese weatherman/woman. You are going to report the weather forecast in Chinese for four major cities in China. The information you are going to report is provided to you in English below. You need to translate the information into Chinese in order to report it. Write your Chinese notes next to the English information.

City 1 上海

It will rain tomorrow morning in Shanghai.

上海________________________________。

City 2 南京(Nánjīng) **Nanjing**

It will rain tomorrow evening in Nanjing.

南京________________________________。

City 3 北京

The sun will come out tomorrow in Beijing.

北京________________________________。

City 4 重庆(Chóngqìng) Chongqing

It will be a cloudy day in Chongqing tomorrow.

重庆________________________________。

V. 沟通练习 Communicative Exercise

你明天要和你的中国朋友们一起旅行。你请你的朋友们为这趟旅行带不同的东西。你担心他们会忘记，所以你想提醒他们。你的信息已用英文写好了。把这些信息翻译成中文，你的中国朋友们才看得懂。

You are going on a road trip with your Chinese friends tomorrow. You have asked several of your friends to bring different things for the trip. You worry that they will forget, so you want to email them to remind them. Your messages are already written in English. Translate them into Chinese so that your Chinese friends will understand.

Message 1

English: Don't forget to bring an umbrella.

Chinese: ________________________________

Message 2

English: Don't forget to bring Chinese music.

Chinese: ________________________________

Message 3

English: Don't forget to bring coke.

Chinese: ________________________________

Message 4

English: Don't forget to bring Chinese books.

Chinese: ________________________________

Message 5

English: Don't forget to bring your dog.
Chinese: ______________________________

第十课 Lesson 10

早餐吃了吗？
Have You Eaten Breakfast Yet?

I. 听力练习 Listening Exercise

李中明正要吃他从家里带来的午餐。张爱林也带了她的午餐来。她刚好在李中明旁边找到了一个座位。听他们的对话并和一位伙伴讨论下列问题。

Li Zhongming is in the cafeteria. He is going to eat the lunch he brought from home. Zhang Ailin also brought her lunch from home. She just found a spot to sit next to Li Zhongming. Listen to the following conversation and discuss the questions below with a partner.

问题 Questions:

1. 李中明吃午餐了吗？
   Has Li Zhongming eaten lunch yet?
2. 张爱林吃午餐了吗？
   Has Zhang Ailin eaten lunch yet?
3. 李中明的午餐有什么？
   What does Li Zhongming have for lunch?
4. 张爱林的午餐有什么？
   What does Zhang Ailin have for lunch?

II. 会话练习 Speaking Exercise

你是一家餐厅的老板。你的餐厅每日提供早、午、晚餐的特色菜。在下面空格画出这些特色菜。每个特色菜一定要有至少一样食物和一种饮料。当你准备好了,两人一组互相问下列问题:(1)早餐有什么?(2)午餐有什么?(3) 晚餐有什么?

You are a restaurant owner. Your restaurant offers daily specials: one for breakfast, one for lunch and one for dinner. Draw an illustration of the special meals in the boxes below. Each of the boxes must have at least one food item and one drink item. When you finish, work in pairs and ask each other three questions: (1)早餐有什么? (2)午餐有什么? (3) 晚餐有什么?

练习结束后,你的老师会请几个同学给大家看他们设计的特色菜。

After you finish the task with a partner, your teacher will select a few students to show their restaurants' daily specials to the class.

早餐

午餐

晚餐

III. 阅读练习 Reading Exercise

陈今和陈丽刚从学校回到家。他们都饿了。读下面陈今、陈丽和他们的妈妈有关晚餐的对话并和一个伙伴回答下列问题。

Chen Jin and Chen Li just arrived home from school. Both of them are hungry. Read the dialogues in the pictures below between Chen Jin, Chen Li and their mother about having dinner and discuss the questions below with a partner.

太棒了！我喜欢吃三明治。
太棒了！我喜欢吃意大利面。

你们想喝什么饮料？
我想喝可乐。
我想喝红茶。

问题 Questions:

1. 今天的晚餐有什么？
   What is there for dinner today?
2. 陈今想吃什么？
   What does Chen Jin like to eat?
3. 陈丽想吃什么？
   What does Chen Li like to eat?
4. 陈今想喝什么？
   What does Chen Jin want to drink?
5. 陈丽想喝什么？
   What does Chen Li want to drink?

IV. 写作练习 Writing Exercise

你的妈妈今天出差，你睡觉前不会回来。用中文写一纸条，问你的妈妈晚餐吃了吗？提醒她要吃晚餐。最后，跟你妈妈说晚安并把你的名字写上。

Your mother is on a business trip and won't be home until after you go to sleep tonight. Write a note in Chinese. Ask your mother, "Have you eaten dinner?", remind her to eat dinner by writing, "Don't forget to eat dinner.", and end the note with, "Good night" and your name at the bottom of the note.

妈妈：

____

____

____

____

V. 沟通练习 Communicative Exercise

你想邀请你中文班的朋友们一起吃午餐。你问他们："我们一起去吃午餐好吗？"如果他们同意，他们会说："好啊。"但是有些朋友忙着学中文没办法跟你去。他们给你的答案可能是：

You want to invite your friends in Chinese class to have lunch with you. You suggest to them, "我们一起去吃午餐好吗？". If your friends agree, they will answer you, "好啊". But some of your friends are busy studying Chinese and can't go with you to lunch. Some of their answers to you may be:

"我现在得写汉字，没有时间吃午餐。"

"我现在得读中文书，没有时间吃午餐。"

"我现在得听中文音乐，没有时间吃午餐。"

"我现在得练习说中文，没有时间吃午餐。"

在教室走动并邀请你的朋友们一起吃午餐。几个人可以跟你一起去？几个人忙着学中文？他们用什么方法学中文？

Walk around the classroom and invite your friends to lunch. How many of them are able to go with you? How many are busy studying Chinese? What kinds of study activities do they need to do?

听力练习文本
Scripts for Listening Exercises

Lesson 1

1. (1) 1. bà　2. pó　3. mù　4. fǎ
5. dú　6. tā　7. nǐ　8. lù
9. gē　10. kū　11. hé　12. jí
13. qǔ　14. xì　15. zhù　16. chī
17. shā　18. rì　19. zì　20. cí
21. sè

(2) 1. pù　2. mó　3. dí　4. mǎ　5. nì
6. gē　7. qú　8. zhè　9. sī　10. cā

(3) 1. tú　2. gǔ　3. kè　4. bó　5. hū　6. nà
7. pà　8. jí　9. lè　10. cā　11. qí　12. sù
13. shì　14. xǐ　15. rè　16. mǐ　17. zū　18. fú
19. zhī　20. dǐ　21. chē

2. (1) 1. zhá　2. fó　3. zè　4. mǐ　5. bù　6. lǜ
7. dǎi　8. něi　9. gào　10. dōu　11. jiā　12. qiáo
13. piě　14. miù　15. guā　16. huó　17. huài　18. duì
19. nüè　20. bān　21. hěn　22. náng　23. děng　24. tòng
25. pián　26. lìn　27. liǎng　28. dīng　29. xióng　30. duàn
31. kǔn　32. zhuāng　33. wēng　34. yuán　35. yùn　36. zhèr

(2) 1. nǚ　2. duō　3. duì　4. zhǎo　5. xiáng
6. zàn　7. hěn　8. níng　9. dùn　10. jiāo

(3) 1. dá　2. bó　3. sè　4. lǐ　5. dù　6. nǚ
7. lái　8. děi　9. pào　10. tōu　11. xiā　12. xiáo
13. biě　14. liù　15. huā　16. guó　17. guài　18. huì
19. lüè　20. sān　21. běn　22. táng　23. děng　24. zòng
25. tián　26. mín　27. niàng　28. tīng　29. qióng　30. tuán
31. kǔn　32. shuāng　33. wēng　34. xuán　35. jùn　36. zhèr

II

1. 1. zān　2. táng　3. xióng　4. jǐ　5. shuāng　6. zhǎo
7. jiě　8. mèi　9. dōu　10. jiù　11. chuǎn　12. qí

2. 1. jiǔ　2. qiú　3. xiū　4. zhuì　5. guì　6. huí
7. chūn　8. shùn　9. lùn　10. jué　11. juān　12. jùn
13. xué　14. xuàn　15. xún　16. zì　17. chī　18. sǐ
19. yào　20. yě　21. yōng　22. wǔ　23. wài　24. wén
25. wáng　26. yuè　27. yuán　28. yǔn

III

1. 1. 请写　2. 走走　3. 想你　4. 你好　5. 你早
6. 我去　7. 很累　8. 请问　9. 总是　10. 也对
2. 1. 不是　2. 不爱　3. 不对　4. 不做　5. 不定
6. 不看　7. 不大　8. 不要　9. 不利　10. 不再
3. 1. 一边　2. 一天　3. 一千　4. 一瓶　5. 一点
6. 一半　7. 一字　8. 一下　9. 一遍　10. 一次

Lesson 3

王老师(男声)：大家好，我是王老师。
陈国(男声)和林丽(女声)：王老师好。
陈国(男声)：王老师，请问您是哪国人？
王老师(男声)：我是中国人。你叫什么名字？
陈国(男声)：我叫陈国。我是美国人。

王老师(男声)：你叫什么名字？
林丽(女声)：我叫林丽。我也是美国人。
王老师(男声)：你是美国哪里人？
林丽(女声)：我是美国德州人。
陈国(男声)：我是美国加州人。

Lesson 4

王红(女声)：我家有三个人。
陈国(男声)：你家有谁？
王红(女声)：我家有爸爸、妈妈和我。你呢？
陈国(男声)：我家有五个人。我家有爸爸、妈妈、两个姐姐和我。
王红(女声)：你家有狗吗？
陈国(男声)：没有，我家没有狗，但是我家有一只猫。你呢？
王红(女声)：我家有三只狗，但是没有猫。

Lesson 5

陈国(男声)：李丽，你喜欢学英文吗？
李丽(女声)：是啊！我喜欢学英文。因为我有很多美国朋友，他们都喜欢说英文。
陈国(男声)：我弟弟和我也喜欢学英文，我们每天写英文字，读英文书。
李丽(女声)：为什么你们喜欢学英文？
陈国(男声)：因为我爸爸是英文老师。
李丽(女声)：我们一起练习英文好吗？
陈国(男声)：好啊！

Lesson 6

李丽(女声)：明天我想和我的中国朋友一起吃中国菜。你想不想一起去？
陈国(男声)：我想和你的中国朋友一起吃中国菜，但是我的中文不好。
李丽(女声)：没关系。我的中国朋友说中文，也说英文。

陈国(男声)：太好了！明天我们一起去吃中国菜。

Lesson 7

李老师(男声)：爱林，长大以后你想做什么？
张爱林(女声)：长大以后我想做中文老师。
李老师(男声)：为什么你想做中文老师？
张爱林(女声)：因为我很喜欢写汉字。我也很喜欢读中文书。
李老师(男声)：太棒了！祝你梦想成真！

Lesson 8

林老师(女声)：你们好。欢迎你们来我家。
王国(男声)，爸爸(男声)，妈妈(女声)：林老师好。
林老师(女声)：请问你们想喝什么饮料？
爸爸(男声)：请问您有什么饮料？
林老师(女声)：我有果汁、汽水、茶、水和咖啡。
爸爸(男声)：我想喝茶。
妈妈(女声)：我也想喝茶。
王国(男声)：我想喝果汁。

Lesson 9

(男声)北京明天上午会出太阳，是晴天。明天下午也是晴天。晚上会下雨，是雨天。北京明天不会下雪。

Lesson 10

李中明(男声)：爱林，你好吗？
张爱林(女声)：我很好。
李中明(男声)：午餐吃了吗？
张爱林(女声)：还没吃呢。你呢？午餐吃了吗？
李中明(男声)：我也还没吃呢。我很饿。

张爱林(女声)：我们一起吃午餐好吗？
李中明(男声)：好啊！
张爱林(女声)：你的午餐有什么？
李中明(男声)：我的午餐有意大利面、青菜和果汁。
张爱林(女声)：我的午餐有炒饭、鸡肉和蛋花汤。

国家级首批一流课程教材　清华大学精品课程教材
纺织服装高等教育"十四五"部委级规划教材
服装史论书系

中外服装史

（简明版）

贾玺增　著

東華大學出版社
·上海·

图书在版编目（CIP）数据

中外服装史：简明版 / 贾玺增著. -- 上海：东华大学出版社，2021.6

ISBN 978-7-5669-1779-9

Ⅰ. ①中… Ⅱ. ①贾… Ⅲ. ①服装—历史—世界 Ⅳ. ① TS941-091

中国版本图书馆 CIP 数据核字 (2021) 第 096752 号

策划编辑：马文娟

责任编辑：季丽华

文字编辑：高路路

装帧设计：上海程远文化传播有限公司

中外服装史（简明版）

ZHONGWAI FUZHUANGSHI JIANMINGBAN

著：贾玺增

出版：东华大学出版社（上海市延安西路1882号，邮政编码：200051）

本社网址：http://dhupress.dhu.edu.cn

天猫旗舰店：http://dhdx.tmall.com

营销中心：021-62373056

印刷：上海盛通时代印刷有限公司

开本：889mm × 1194mm 1/16

印张：8

字数：195千字

版次：2021年6月第1版

印次：2024年1月第2次

书号：ISBN 978-7-5669-1779-9

定价：58.00元

序

服装史是高等院校服装专业教育的必修课程，学习服装史不仅为学生提供知识储备和设计素材，也对提升他们的创新力、提案力和审美力有很大帮助。学习服装史对服装专业人才培养具有重要意义。

自2015年开始，清华大学美术学院贾玺增副教授撰写的“十三五部委级规划教材•服装史论书系”陆续面世，现已出版了《中外服装史》（国家精品课程教材、清华大学优秀教材）和《中国服装史》（国家精品课程教材）两本，还有一本《西方服装史》即将出版。这三本教材阐述系统、内容详实、资料丰富，书中穿插大量当代复原作品和运用历史元素设计的时装案例，文字和图片的信息总量较大。因此，在阅读时需要读者静下心来，用较长的时间认真学习，循序渐进，才能慢慢掌握和理解书中的内容。

“十三五部委级规划教材•服装史论书系”得到了业内专家、中国高等院校服装专业师生和广大读者的一致好评，尤其是《中外服装史》已被国内120余所院校作为课程教材和研究生入学考试教材。有一些读者询问能否再出版一套相对轻松、简单的入门级图书。鉴于这个目的，贾玺增副教授又撰写了这三本“简明版”系列丛书，包括《中国服装史（简明版）》《西方服装史（简明版）》《中外服装史（简明版）》，以此作为“十三五部委级规划教材•服装史论书系”丛书的扩展和补充，使广大读者、高等院校服装类专业学生更便捷地学习服装史课程和知识。

在撰写的过程中，贾玺增副教授遵循易读、易学、易掌握的原则，将服装史里的知识点提炼出来，经过谨慎选择，以词条的形式撰写（尽可能减少古代文献），以便于读者阅读、掌握和记忆。本系列丛书还注重正文、图片和图注的紧密结合和编排，使读者能够在短时间内通览全书，快速建立服装史的基础知识和理论体系，为今后的深入学习打下基础、创造条件，并保持轻松的学习状态和良好的学习兴趣。

贾玺增副教授长期从事服装史学术研究，在设计实践和课堂教学等领域取得了显著成绩。他参与策划的“‘华夏衣裳’中国高等院校服装史教学与学术研讨会”（中国纺织服装教育学会、清华大学美术学院主办），吸引近百所院校参与，旁听观众几十万人。此外，贾玺增副教授还连续多年在东华大学出版社举办“全国服装专业教师暑期高级研修班”，为全国各地的服装专业老师进行授课，分享教学经验和学术成果，在服装史教学方面起到了很好的“传帮带”作用，同时也推动了服装史课程教学水平的提高。借此机会，对贾老师的辛勤工作表示感谢！

五千年来中华文明生生不息，服装是重要的传承载体，“一丝一缕”中歌颂着先辈们对美好生活的向往和追求。相信本套丛书一定会受到读者的欢迎和喜爱。

中国纺织服装教育学会副会长、秘书长

2021年2月25日

目 录

第一章　公元前 1 万年—公元前 5 千年

在人类文明尚未开化的远古时期，工具简单粗糙，生活环境险恶，人类的服装意识还不明确。旧石器时代初期，仅局限于用兽皮、树叶等原始材料对人体的局部进行包裹，以达到遮蔽身体和御寒的目的。至旧石器时代晚期，华夏先民开始了相对稳定的采集、渔猎和洞穴生活，石器的加工比前一阶段更精确、规整和美观，形状也趋于定型化、小型化、多样化。

石器

在距今约 300 万至 1 万年前的旧石器时代，人类主要使用打制、砍砸而成的石制器工具（图 1–1）。掌握工具制造技术是人类进化史上的一个重要标志。随着生产工具的革新，产生了以“磨光”替代“砍砸”的工艺。这种经过磨光、带有刃口的工具，比旧石器时代晚期的工具更进步，人们称之为“新石器”。

磨光后的石器（图 1–2、图 1–3）形制小巧，造型悦目，令人爱不释手。大汶口文化遗址、良渚文化遗址出土了大量新时器时代晚期的石制、玉制等首饰。如上海青浦出土的良渚文化玉项链（图 1–4）和浙江余杭瑶山墓出土的良渚文化玉串饰（图 1–5）。

石器时代晚期，一种更锋利、更实用的石钺开始出现。除了实战功能外，石钺还是军事控制权的象征。当时的部落首领，集军权、神权于一身，石钺则是这些部落首领手中的“权杖”（图 1–6）。

图 1–1　双面砍砸器（大英博物馆藏）

图 1–2　新石器时代石斧

图 1–3　新石器时代石锛

图 1–4　良渚文化玉项链（上海青浦出土）

前者由大小不等的腰鼓形和圆珠形穿孔玉珠穿成。最下一粒玉坠呈铃形，柄部有一穿孔。坠的两侧为玉管，浮雕双目和嘴组成兽面纹。

图 1–5　良渚文化玉串饰（浙江余杭瑶山墓出土）

图 1–6　礼用石钺（安徽含山凌家滩遗址出土）

兽齿

在旧石器时代晚期，石制工具已趋于定型化、小型化。直立行走，以及火和盐的使用，促使中国古人的智力进一步发展。华夏祖先使用磨制和穿孔技术，将骨、角、牙等天然材料加工成为具有美化功能的饰品。北京山顶洞人、辽宁海城小孤山等旧石器时代遗址中都有用贝壳（图 1–7）、兽牙（图 1–8）、石珠和鸟骨做成的穿孔饰品出土。最引人注目的是四川广汉三星堆二号祭祀坑出土的 3 枚虎牙（图 1–9），其钻孔、磨制和染色技术，显示了旧石器时代晚期人类祖先生产技能的提高和审美意识的出现。这些饰品不仅是美化人类自身的一种器物，还是权力、意志、力量、勇敢和财富的标志。

图 1–7　穿孔饰品（北京山顶洞人遗址出土）

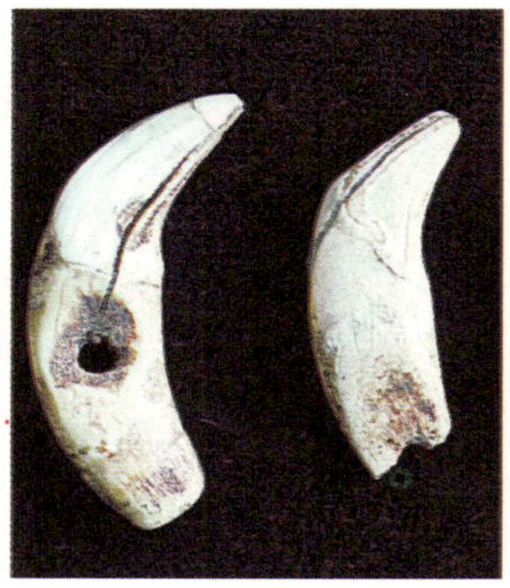

图 1–8　穿孔兽牙（北京山顶洞人遗址出土）

长度为 9.3 ~ 11.3 厘米。由于长时间与青铜器埋藏在一起，虎牙被铜锈侵蚀呈碧绿色。每个虎牙根部都有穿孔，可能是作穿戴系挂的装饰之用。

图 1–9　3 枚虎牙（四川广汉三星堆遗址出土）

由 8721 颗细骨珠制成。估计其全长可达 16 米左右，可反复来回围绕颈、臂及腰部。

图 1–10　细骨珠串饰（陕西临潼姜寨少女墓出土）

长 27.3 厘米，宽 3.8 厘米。一面以甲骨文刻载帝辛将猎获的犀牛赏赐宰丰之事（有学者认为该刻字为后人仿制），另一面刻兽面、蝉纹和虺龙纹，并嵌有绿松石。

图 1–11　商代宰丰骨匕（传河南安阳出土，中国国家博物馆藏）

杯身中腰微束，一侧有与杯身等高夔龙形把手。上下口有两条素地宽边，绿松石的条带间隔为四段：第一段为饕餮纹三组，两侧有身有尾，眼、眉、鼻镶嵌绿松石；中间有两组饕餮纹，兽口下面为一个大三角纹，三角纹两侧有对称的夔纹，头朝下，尾向上，饕餮的口、眼、鼻及三角纹都镶嵌绿松石；第三段刻三个变形夔纹，眼部镶嵌绿松石；第四段的三组饕餮纹眼、鼻同样镶嵌绿松石。

图 1–12　嵌绿松石象牙杯（河南安阳殷墟妇好墓出土）

骨器

在石器和青铜时代，骨器在中国先民的生产和社会活动中占有重要地位。畜牧产品是人们赖以生存的重要生活来源，动物骨骼为骨珠、骨刀、骨匕、骨铲、骨锥、骨镞、骨针、骨梳及骨纺轮等骨器的制造提供了充足的原料。其实物如 1972 年陕西临潼姜寨少女墓出土的一串细骨珠串饰（图 1–10）、中国国家博物馆藏传河南安阳出土的商代宰丰骨匕（图 1–11）、河南安阳殷墟妇好墓出土的嵌绿松石象牙杯（图 1–12）。

长8.2厘米，一端尖锐，另一端有直径0.1厘米的针孔。其后，又有一些骨针被陆续发现，如西安半坡新石器时代遗址出土了328枚骨针，最小的仅长9厘米，直径0.2厘米，针孔直径0.1厘米，与今天大号钢针差不多。

图1-13　骨针（浙江余姚河姆渡遗址出土）

羊毛毡裁制成风兜状，帽身连缀耳护，下为敞口。耳护下垂至颌部，原本应缀有绳带。整件毡帽呈灰白色，沿口用等长的粗线缝缉，出土时戴在女性尸体头部。

图1-16　毡帽实物（新疆楼兰罗布泊孔雀河北岸古墓沟出土）

图1-14　原始社会晚期皮衣（新疆哈密古墓出土）

本白色羊毛上缀着红色线绳，左侧缀白鼬皮，鼬头悬在帽子前部，帽子上还绑有羽饰，是用赤色的毛线绑在木桩上，然后插在帽子上的。

图1-17　白鼬皮饰羽毡帽（新疆罗布泊楼兰遗址西汉末或东汉初期古墓中出土）

此鞋称“裘茹克”，根据脚的大小切割兽皮，再打小洞穿绳收紧便可。冬天毛在里，夏天毛在外。“裹脚皮”的使用，使人类摆脱了跣足行走的困境，加快了行走的步伐，扩大了活动范围，有力地促进了农耕渔猎等活动的发展。

图1-15　棕色毛皮鞋（新疆罗布泊铁板河出土）

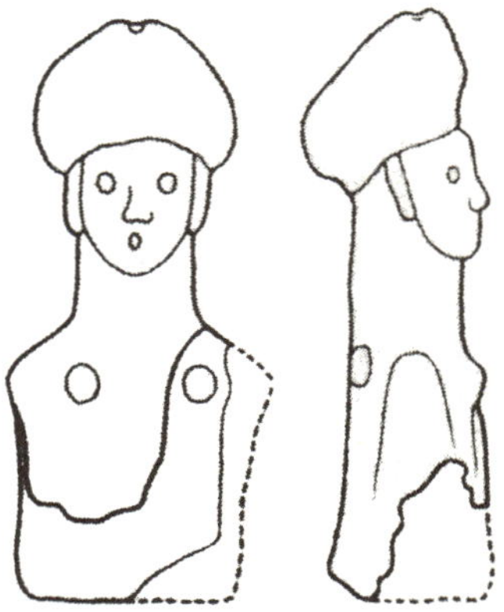

图1-18　陶塑人像（湖北天门邓家湾石家河文化遗址出土）

皮衣

在纺织品尚未发明前，兽皮是中国先民最易获取的制衣材料，即《礼记·礼运篇》所载“昔者……未有丝麻，衣其羽皮。”从考古实物分析，华夏祖先很可能在距今两万年前就已掌握了使用骨针（图1-13）缝制兽皮的技术。缝线由动物韧带纤维或植物韧皮纤维搓捻而成。兽皮衣制好后，华夏祖先还要用咬嚼或揉搓的方法进行皮革的初步鞣制，使其成为穿用更为舒适的柔软材质（图1-14、图1-15）。

毛毡

除了兽皮，中国古人还以毛毡制做衣帽（图1-16、图1-17）。制毡是毛纺织的前导。羊毛毡历史悠久，在华夏大地存在了很长时间，而且也有着很好的发展。中国新疆维吾尔自治区、陕西省、甘肃省等地都有长期使用羊毛毡用品的历史。华夏祖先戴毡帽的形象如湖北天门邓家湾石家河文化遗址出土的头戴筒形平顶窄檐帽的陶塑人像（图1-18）。

第二章　公元前 5 千年—公元前 2 世纪

第一节　中国：夏商周

一、商代

商朝约为公元前 1600 至公元前 1046 年，是中国第一个有直接同期文字记载的王朝。夏朝诸侯国商部落首领商汤，率诸侯国于鸣条之战灭夏后在亳（今商丘）建立商朝。之后，商朝国都频繁迁移，至其后裔盘庚迁殷（今安阳）后，国都才稳定下来，在殷建都达 273 年，所以商朝又称“殷”或“殷商”。商朝延续了 600 年左右，末代君王商纣王于牧野之战中被周武王击败后自焚而亡。

随着生产水平的提高，社会贫富分化初显，手工业逐步从农业中分离出来，服饰也从遮蔽、保暖的物质需求，逐步向材料、质地、文化、等级和形制等方向发展。当束发在华夏先民中普遍流行时，发笄还成为了头部的装饰和身份的象征（图 2–1、图 2–2）。

从出土的玉雕、石雕和陶俑等文物分析，当时社会上层贵族服饰的上身一般为交领右衽、窄袖及腕、织绣华丽纹饰（如兽面纹、矩形纹、双钩云纹等）的上衣，腰束宽带，下身穿有纹样装饰的裳或裤。其服饰可通过以下 4 件雕像得到大致的了解：

河南安阳殷墟妇好墓出土的戴箍冠圆雕玉人像（图 2–3），头编一长辫，戴一头箍前有横式筒状卷饰，人像穿交领窄长袖衣，衣长及足踝，束宽腰带，上饰格菱纹，左腰插一卷云形宽柄器，着鞋；河南安阳殷墟遗址出土的大理石圆雕右半身残人像（图 2–4），上衣盖臀，交领右衽，腰束宽带，胫扎裹腿，足穿翘尖之鞋；传河南安阳殷墓出土的圆雕石人立像（图 2–5），石人双手拱置腰前，上衣为交领右衽、窄袖，下裳前摆过膝，后裾齐足，足着平底无跟圆口屦，腹下悬斧形“蔽膝”；四川广汉三星堆遗址出土的大型青铜立人像（图 2–6），头戴花形冠，外衣饰蚕纹、龙纹、尖角纹、饕餮纹等，上衣佩组带，下裳分前后两片，后片长至足踝，呈燕尾状，前片略短，露出小腿部。

图 2–1　菱形头骨笄（河南安阳殷商妇好墓出土）

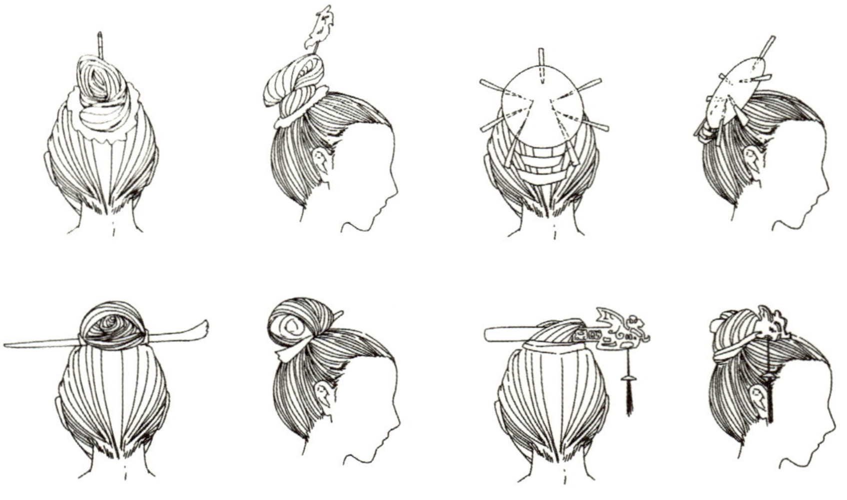

图 2–2　商代女性发式及发笄簪插方式示意图

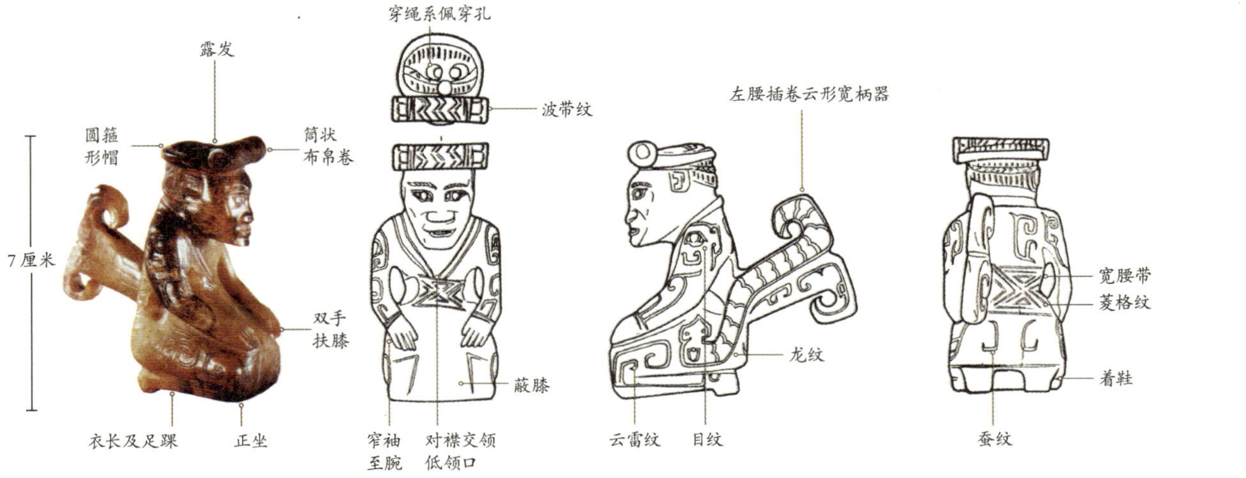

图 2-3　戴箍冠圆雕玉人像（河南安阳殷墟妇好墓出土）

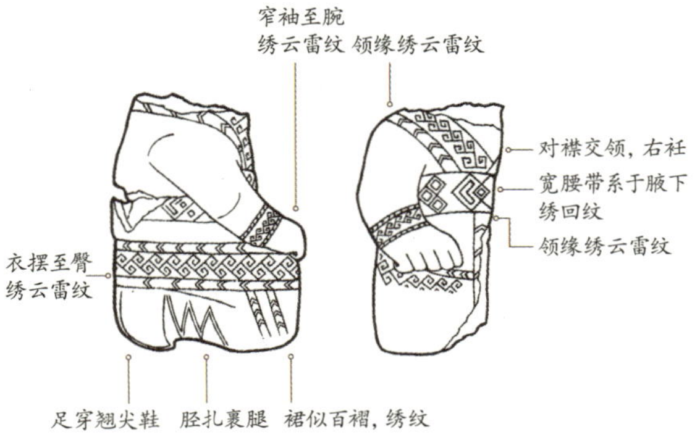

图 2-4　大理石圆雕人像的右半身残像（河南安阳殷墟遗址出土）

图 2-5　圆雕石人立像（传河南安阳殷墟墓出土，哈佛大学福格美术馆藏）

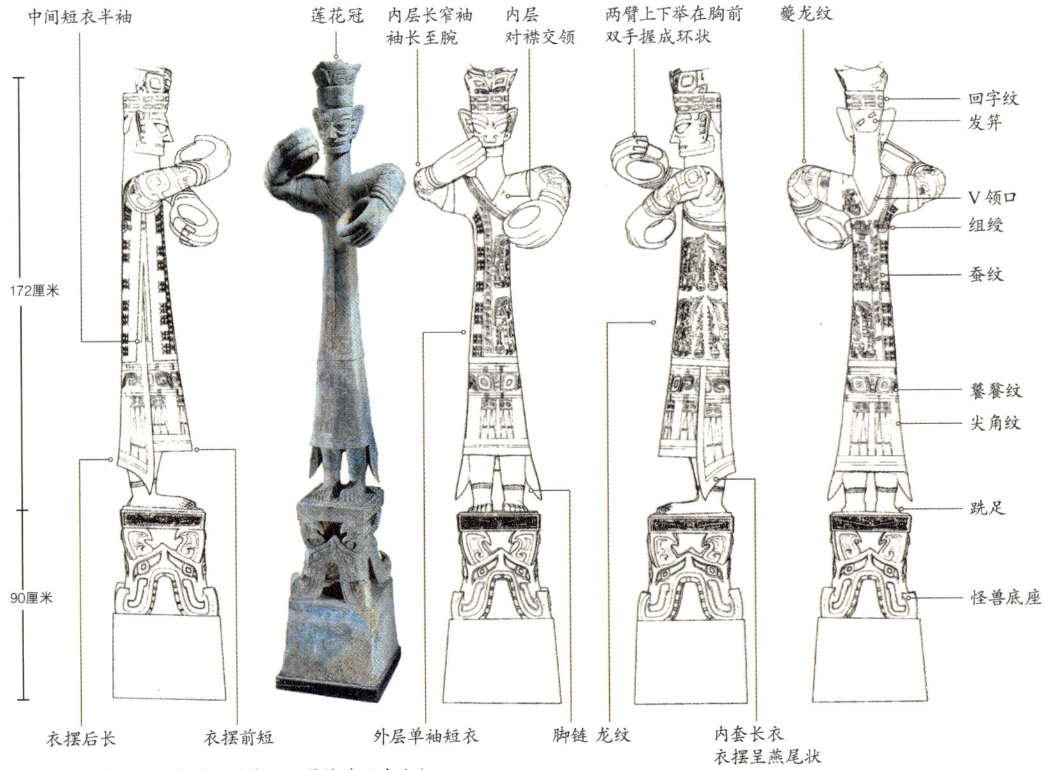

图 2-6　大型青铜立人像（四川广汉三星堆遗址出土）

二、周代

西周：公元前 1046—公元前 771 年

东周：公元前 770—公元前 256 年

周朝存在约 800 年，从公元前 1046 年到公元前 256 年，共传 30 代 37 王，分为西周和东周两个时期，为中国历代最长的王朝。西周建都镐京（今陕西西安附近），到公元前 771 年结束。

周代政治制度是以血缘关系为基础的封建宗法制度，商人尚鬼神、奢靡、酗酒的风气已被改变。周王朝为了维护统治，推行了一整套包括冕服在内的维护统治阶级内部秩序的礼乐制度。冕服是中国古代等级最高、使用最久、影响最广泛、最具代表性的男子礼服(图 2–7)。它主要由冕冠、上衣、下裳、舄等主体部分以及蔽膝、绶、带、佩等其他配件构成。冕服之制，传说殷商时期已有，至周定制规范，趋于完善，汉代以降，历代沿袭（图 2–8）。至清朝建立，废除汉族衣冠，冕服制度在中国随之终结。历史上除中国外，冕服在东亚地区的日本、朝鲜、越南等汉字文化圈国家中亦曾作为国君、储君等人的最高等级礼服。

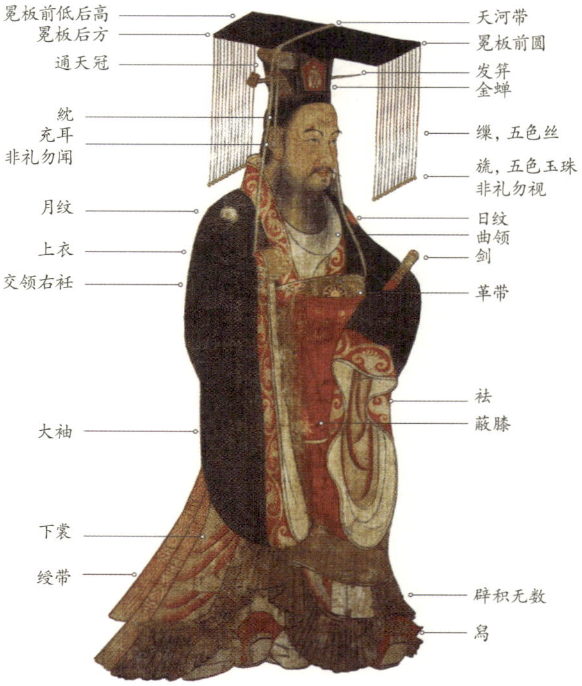

图 2–7 《历代帝王图》中身穿冕服的古代帝王像

图 2–8 南宋《孝经图卷》中帝王百官穿冕服的场景

六冕

大裘冕，祭天时所穿以大裘为外衣的冕服，因祭天而称“大”；衮冕，祭先王时所穿以卷龙为首章的冕服；鷩冕，祭祀先王、飨射典礼时所穿以华虫为首章的冕服；毳冕，遥祭山川时所穿以有毳毛的虎蜼为首章的冕服；絺冕，小祀时所穿以粉米为首章的冕服；玄冕，祭社稷时所穿上衣无文（仅下裳施黻一章）的冕服。衮冕、鷩冕、毳冕、絺冕皆因首章得名，玄冕因上衣无章纹，随衣之玄色而得名。

被许可穿冕服的身份有天子、公、侯、伯、子、男、孤、卿、大夫等（图 2–9）。其中，唯有天子得以穿全六冕；公之服，衮冕以下；侯、伯之服，鷩冕以下；子、男之服，毳冕以下；孤之服，絺冕以下；卿、大夫之服，只有玄冕。

图 2–9　宋代《三礼图》中的六冕形象

图 2–10　冕冠（山东邹城明鲁荒王朱檀墓出土）

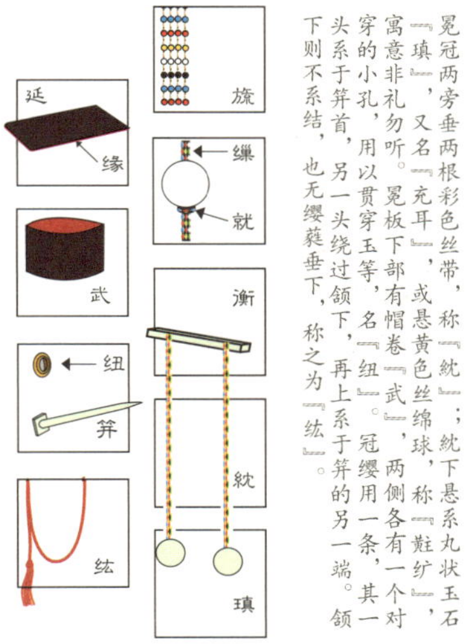

图 2–11　汉代冕冠示意图（贾玺增绘制）

冕冠

冕服因冕得名，冕冠的顶端有一长方形的冕板，被称为“延”或“綖”。其形状为前圆后方，象征天圆地方；其色为“上玄下纁”，以应天地之色。冕板前低后高，为“俛”形，寓意谦逊，故名冕冠；前后垂挂赤、青、黄、白、黑五色玉珠，称“旒”，寓意非礼勿视；穿旒的丝绳以五彩丝线编织，谓之“缫”或“藻”。冕旒数量是身份标志，十二旒为帝王专用（共 288 颗），以下分等级递减，诸侯九旒，上大夫七旒，下大夫五旒，士三旒。山东邹城明鲁荒王朱檀墓出土的冕冠（图 2–10）上悬挂九缫九旒，共 162 颗旒。汉代冕冠具体部位名称如笔者绘制的示意图（图 2–11）。

上衣

冕服上衣式样为交领右衽、博衣大袖。交领右衽是汉族服饰的基本特征和明显标志。其外观为左襟在上并压住右襟，使左右两衣领形成交叉状。其外观如英文字母“y”（图2-12）。汉服右衽衣襟同时体现着阴阳生死的说法。衣襟向右掩视为阳，表示生的人；衣襟向左掩视为阴，表示故去的人。

下裳

冕服下裳从远古人类的遮羞布演变而来。其形制特点有二：第一，裳为前后两片，这是男裳与女裙（一片式）的主要区别；第二，裳上有褶裥（辟积）无数（图2-13）。

图2-12　明代《中东宫冠服》中的冕服上衣

图2-13　明代《中东宫冠服》中的冕服下裳

图2-14　明代《中东宫冠服》中的冕服蔽膝

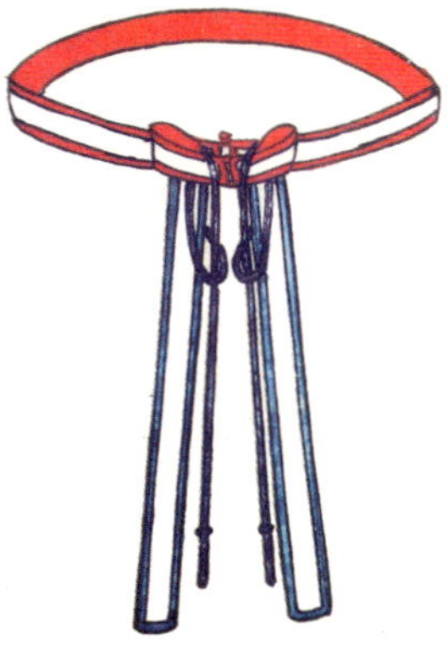

图2-15　明代《中东宫冠服》中的大带

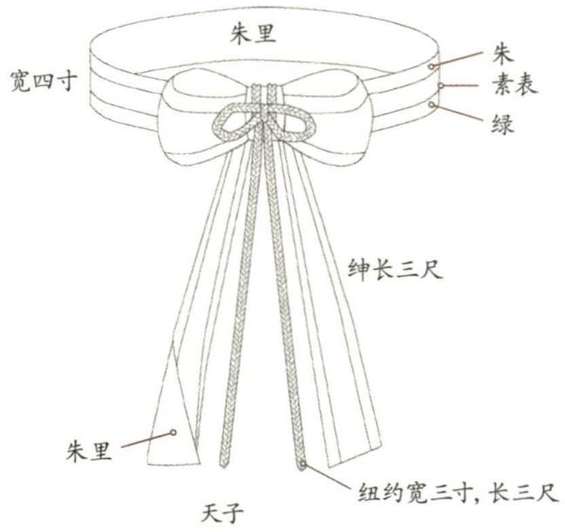

图2-16　周代天子大带示意图

蔽膝

蔽膝是华夏祖先遮羞物的遗制，以熟皮或绢为之，意在尊祖。蔽膝在朝服中称为“韠”，在冕服中称韍、铍和芾（图2-14）。根据古代的礼制，大夫以上才佩赤带。天子用直，朱色，绘龙、火、山三章。上系于革带，悬垂于膝前。与女裙不同，蔽膝稍窄，且要长到能遮盖膝盖的部位，用在衣裳制礼服上要求与帷裳下缘齐平。在殷商时期出土的圆雕石人立像中，小腹之下的下裳前，多附一下广上狭形如斧形的蔽膝。东汉郑玄释其形曰“圆杀其下”。

大带

冕服所用腰带为大带（图2-15、图2-16）和革带。大带的功能是以长短、颜色区分系者的身份（大夫大带，宽为四寸；士之大带，宽为二寸；天子大带为素，朱色衬里，缘饰上以朱，下以绿）。大带的系束方式是由后绕向前，于腰前缚结。带子前面垂下的部分称“绅”，有“敬谨自约”之意，故大带又称“绅带”。革带是以皮革制成的腰带，古称“鞶带”或“鞶革”，作用是佩挂蔽膝，佩玉、绶和剑等。

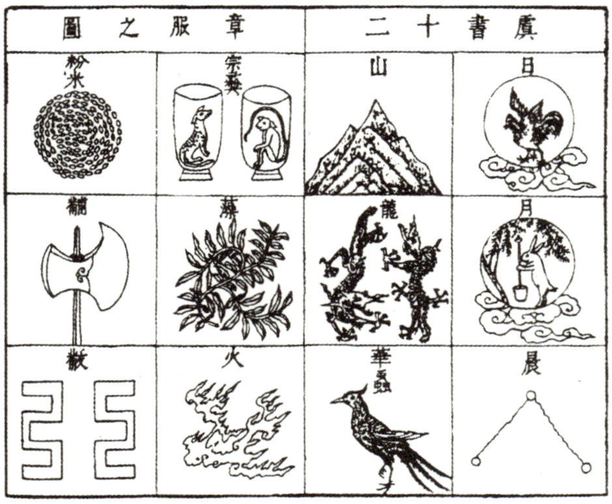

图 2-17　宋代《三礼图》中的十二章纹

图 2-18　清代龙袍上的十二章纹

十二章纹

“十二章纹”，即以十二种纹饰（取天数十二），或画、织、绣在冕服上。不同身份的人所施章纹数各有不同，即天子全备十二章；公用山以下之九章；侯、伯用华虫以下之七章；子、男用藻以下五章；卿大夫用粉米以下三章（图 2-17、图 2-18），章纹与冕服等级参见表 2-1。

日、月、星辰：取其照临，代表三光照耀，象征着帝王皇恩浩荡，普照四方。日、月、星辰被分别施于帝王冕服的两肩和领被。因其代表的是天相，故为“天子”，即皇帝独用；山，取其稳重，代表着稳重性格，象征帝王能治理四方水土；华虫，即雉鸡，是五彩的鸟，取其文理，象征王者要“文采昭著”；火，取其光明，象征帝王处理政务光明磊落；粉米，取其滋养，象征着皇帝给养着人民，安邦治国，重视农桑；藻，水藻，是隐花植物类，没有根、茎、叶等部分的区别，取其洁净，象征皇帝的品行冰清玉洁；龙，取其应变，是一种神兽，变化多端，象征帝王们善于审时度势地处理国家大事和对人民的教诲；宗彝，取虎之威猛，和蜼之有智，遂有避不祥之意，亦表示不忘祖先；黼，左白右黑的斧行图案，取其决断，象征皇帝做事干练果敢、明断是非；黻，由两个半黑半青之色的相背“己”字组成，取其明辨，代表帝王能明辨是非、知错就改、背恶向善的美德，隐含着君臣离合之意。黻纹在十二章纹中排位最低，因此用于衣领，这是人们最易见到的地方。

表 2-1　冕服之章目及章次（郑玄注）

种类	总章数	衣			裳		
衮冕	九章	五章	龙、山、华虫、火、宗彝	画	四章	藻、粉米、黼、黻	绣
鷩冕	七章	三章	华虫、火、宗彝		四章	藻、粉米、黼、黻	
毳冕	五章	三章	宗彝、藻、粉米		二章	黼、黻	
絺冕	三章	一章	粉米	刺绣	二章	黼、黻	
玄冕	一章	×	×	×	一章	黻	

冕服上十二章纹的使用反映了中国古人“万物服体”的思想观念。远古的原始社会在未产生掌管一切的至上神时，总是把这些自然物和自然力看成有生命、意志和伟大力量的对象而加以崇拜。

君士四弁

祭祀场合之外一般穿着以弁为冠、上衣下裳制的礼服（图 2–19）。其穿用等级仅次于冕服。《周礼·春官·司服》载："凡兵事，韦弁服；眡朝，则皮弁服。凡甸，冠弁服。凡凶事，服弁服。"爵弁是身份比冕服等级略低的礼冠，即《周礼·春官·司服》所载"凡服尊卑之次，系于冠，冕服为上，弁服次之，冠服为下。"爵弁服是古代士助君祭时的服装，也是士的最高等的服装。皮弁服（图 2–20、2–21）是周代天子视朝、郊天、巡牲等场合穿的服装，又为诸侯朝见帝王、视朔及田猎时穿的服装。韦弁服是周天子军事时穿用的戎服（图 2–22）。冠弁服是周代天子田猎之服，古以田猎习兵事。

图 2–19　宋代《三礼图》中的爵弁服、皮弁服、韦弁服、冠弁服

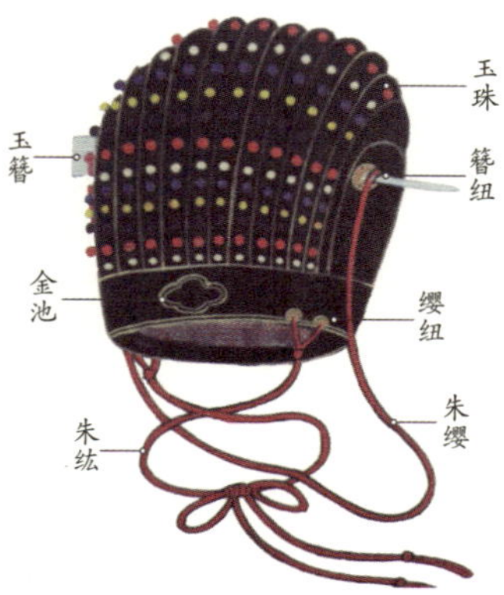

图 2–20　明代《中东宫冠服》中的皮弁

图 2–21　《历代帝王图》中的南朝末陈后主画像

图 2–22　《历代帝王图》中的隋炀帝画像

命妇六服

周代王后及命妇的礼服（朝服和祭服）与帝王和群臣的六种冕服、四种弁服相对应的是六服制度。《周礼·天官·内司服》载：“掌王后之六服：袆衣、揄狄（翟）、阙狄（翟）、鞠衣、襢衣、禒衣。”其中，袆衣、揄狄、阙狄统称“三翟”，为祭服；鞠衣、襢衣、禒衣统称“三衣”，为公服和常服。均剪帛为翟（野鸡）形，上施彩绘，缝缀于衣以为纹饰。此外，六服衣式均为上下连属的一部式服装，喻女德专一。夹里衬以白色纱縠，以便显示出衣纹色彩。帝后六服皆为丝质，仅在衣的用色和纹饰上有所区别（图2–23）。

在中国传统文化中，玄色为天色，其色至尊，故以袆衣为玄色。余下五服色以青、赤、黄、白、黑为序，是根据五行相生之色。

除了服饰，周代后妃及命妇发式和装饰品亦相当丰富。《周礼·天官》载有“追师：掌王后之首服。为副、编、次、追、衡、笄。”其中以“副”为最盛饰，“编”和“次”次之。所谓“副”是在头上加戴假发和全副华丽的首饰，“编”是在加戴假发的基础上加一些首饰，“次”是把原有的头发梳编打扮使之美化。“追”是动词，“衡”和“笄”是约发用的饰品，“追衡、笄”是指在头发上插上约发用的衡和笄。也有人把“追”解释为玉石饰物，“衡”是悬于耳朵两旁的饰物，“笄”贯于发髻之中（图2–24）。后及内外命妇的足服：凡祭服皆为舄，余服以屦，色皆随裳色。

图2–23　河南打虎亭汉墓壁画中头戴大手髻、步摇、花钿的女性形象

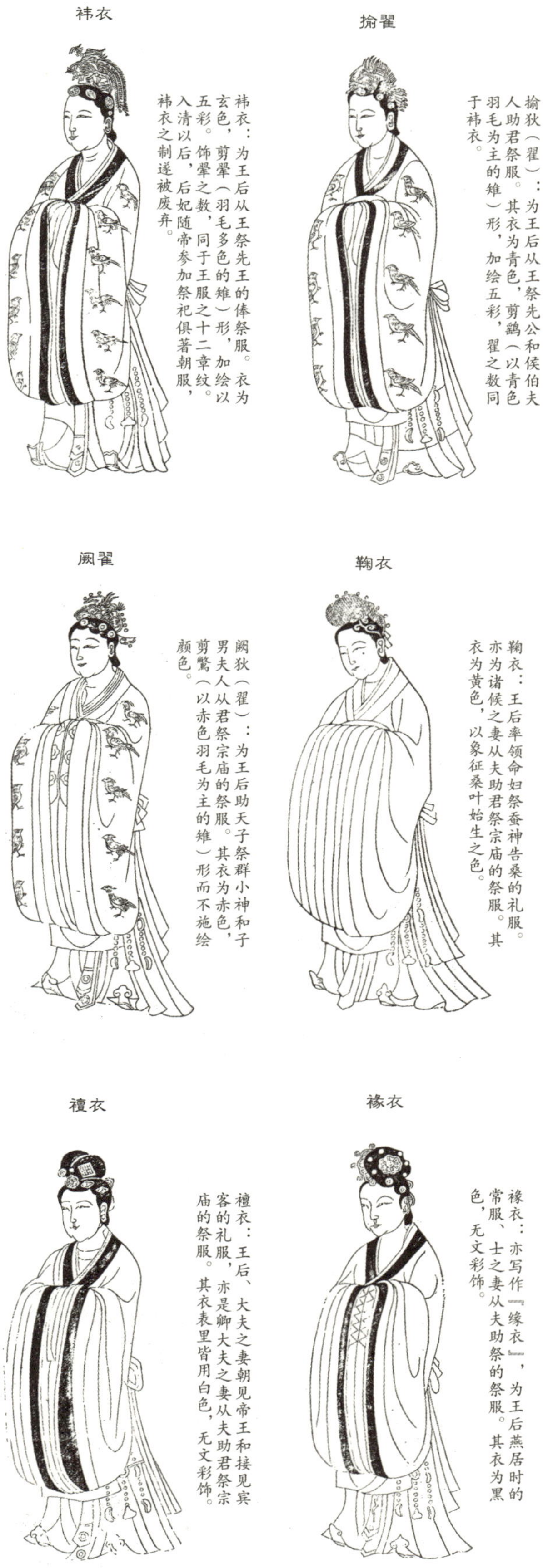

图2–24　宋代《三礼图》中的王后六服

玄端

玄端，亦称元端，为周代的法制服装（图2-25）。上自天子，下至士人，皆可服之。天子服之以燕居；诸侯则以此服祭宗庙；大夫、士朝飨、入庙、朝见父母时服玄端（夕祭则服深衣）。玄端的剪裁方正平直，又黑色无纹饰（同皮弁服），故名。其实物如《历代帝王图》中隋炀帝穿的黑色上衣（见图2-22）。

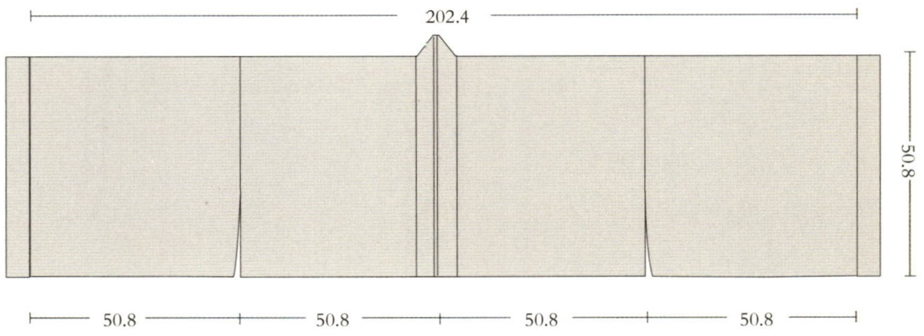

与玄端相配的首服是“委貌冠”，其形制和冠饰同皮弁类似，只是不用白鹿皮，而用玄色缯、绢制成。

图2-25 玄端款式图（贾玺增绘制）（单位：厘米）

深衣

春秋战国时期，人们穿用最多，流行最广泛的服装是深衣。深衣为士以上阶层的常服，士人的吉服（较朝服、祭服次等服饰），庶人的祭服。深衣长度在足踝间，所谓“长毋被土”。

据儒家典籍记载，中国古人出于尊古和文化象征的需要，创制了上下分裁的一体式深衣，还从文化、伦理的角度赋予了它公平（下摆齐平如秤锤和秤杆，象征公平）、正直（衣背中缝线垂直，象征正直）、礼让（衣袖作圆形以与圆规相应，象征举手行揖）、无私（衣领如同矩形与正方相应，象征公正无私）的意义，甚至将天地乾坤、日月轮回（下裳用六幅，共裁十二片，以应一年十二月）的诸多象征也纳入其中（图2-26～图2-29）。

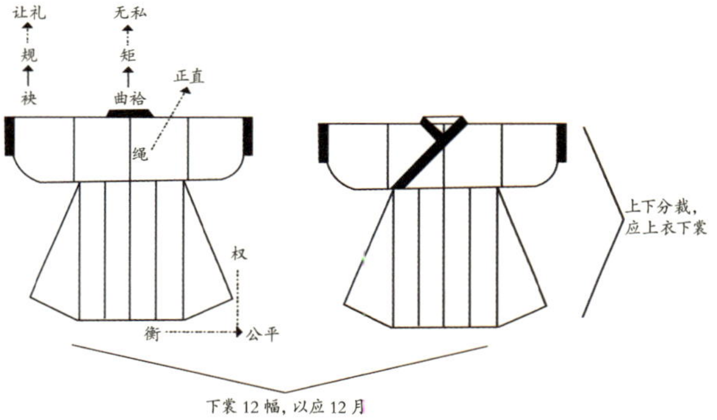

图2-29 深衣中的文化象征（贾玺增绘制）

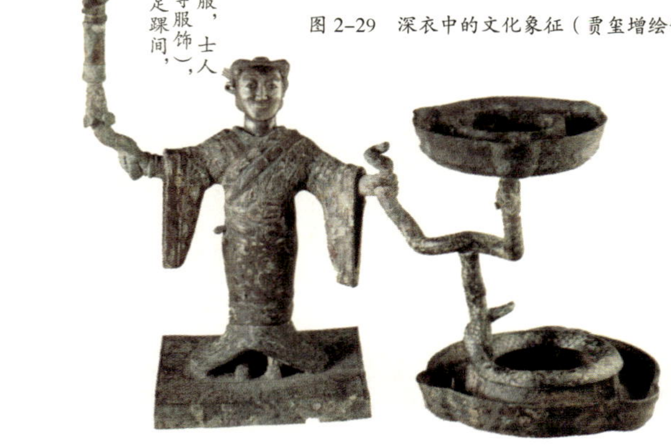

深衣为士以上阶层的常服，士人的吉服（较朝服、祭服次等服饰），庶人的祭服。深衣长度在足踝间，所谓“长毋被土”。

图2-26 战国铜人灯座（河北平山出土）

图2-27 带钩执烛青铜人像（河北易县燕下都遗址出土）

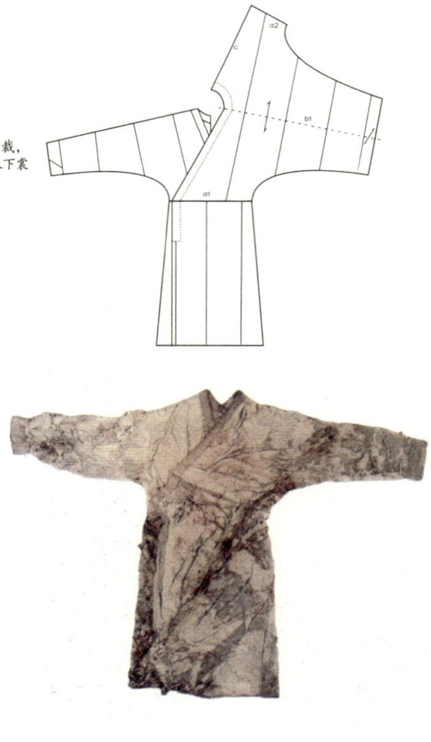

图2-28 素纱绵袍N1和裁剪示意图（湖北江陵马山楚墓出土）

冠礼

随着周代礼仪制度的不断完善，首服的社会功能被不断强化。据《仪礼·士冠礼》和《礼记·冠义》记载，贵族男子到了 20 岁时，需举行象征成年的冠礼（图 2-30）。冠礼仪程由筮卜吉日开始，定了日期后，冠者的父兄须邀请来宾作为青年成人的见证者。冠礼由冠礼者的父亲主持，由指定的贵宾为其加冠三次。初加缁布冠（图 2-31），象征将涉入治理人事的事务，即拥有人治权；再加皮弁（图 2-32），象征将介入兵事，拥有兵权，所以加皮弁的同时往往佩剑；三加爵弁（图 2-33），拥有祭祀权，即为社会地位的最高层次。

图 2-30　电视剧《秀丽江山之长歌行》中戴爵弁的画面

笄礼

据《仪礼》记载，周代女子年满 15 岁就被看作成人。在此之前，她们的发式大多为丫髻，还没有插笄的必要。到 15 岁时，如果已经许嫁，便可如成年女子一样梳发挽髻了。这时就需要使用发笄，故此，古时称女孩成年为“及笄”。

如果女孩没有许嫁，到 20 岁时也要举行笄礼，由一个妇人给及龄女子梳一个发髻，插上一支笄，礼后再取下。周代女子绾髻插笄的样子如河南光山春秋墓女墓主发髻上所插两支木笄（图 2-34）。

冠者换玄端服，分别为：缁布衣（黑色）、玄裳、爵韠。

图 2-31　缁布冠

冠者服白色裳、白色蔽膝。

图 2-32　皮弁

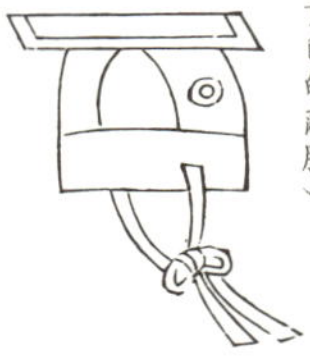

冠者服纯衣（丝衣）、纁裳（浅红色的裳）、韎韐（赤黄色的蔽膝）。

图 2-33　爵弁

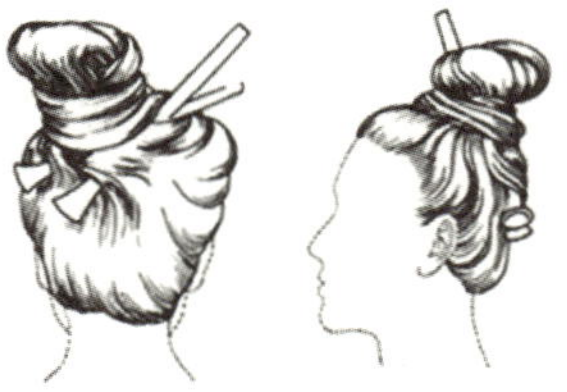

图 2-34　河南光山春秋墓女墓主发髻上斜插两根木笄

佩玉

礼贯穿于周代社会的各个方面，而玉成为实现礼的一个重要手段。当时贵族服装上都有玉饰，所谓“君子佩玉”，玉饰是贵族身份不可缺少的标识。如湖北江陵马山一号楚墓出土的木俑身上佩挂了很长的组玉佩（图 2-35）。其实物如山西曲沃北赵村 92 号墓出土的西周组玉佩（图 2-36）。

《礼记·聘义》中把玉概括为仁、智、义、礼、乐、忠、信、天、地、德、道，共十一德。通过对玉自然属性的分析，把玉的表象和本质特征与儒家道德观念紧密结合在一起，从而阐明了玉器人格化的概念。

组玉的佩戴还有为君子“节步”之用。君子行走，步子不能太快，不能太慢，要快慢适度，迟速有节，从而使得组玉佩发出愉悦的鸣响；如果君子行步匆忙，步履慌乱，组玉佩则会相互碰撞而发出杂音。

图 2-35　木俑（湖北江陵马山一号楚墓出土）

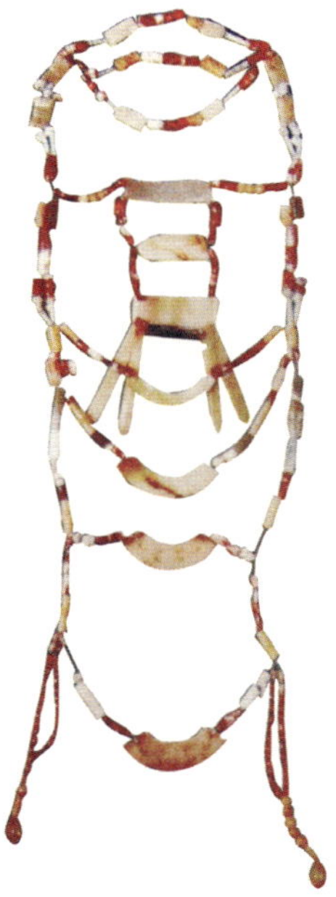

西周组玉佩主要由珩、璜、冲牙、瑀、琚、珠组成。其上为磬状“珩”，两侧系组绶，各垂半璧形而内向的“璜”。中组贯一球状的“瑀”，下垂新月形俯状的“冲牙”。西周和东周两个时期的组玉佩中主要构件也不同，西周是璜，东周为珩。

图 2-36　西周组玉佩（山西曲沃北赵村 92 号墓出土）

第二节　西方：古埃及、美索不达米亚、爱琴文明、古希腊

公元前 3500 年开始的尼罗河古埃及文明，和同时期的两河流域苏美尔文明一起，形成了人类历史上最古老的两大文明摇篮。此后 1000 多年出现了爱琴文明——希腊文明和印度河文明。

一、古埃及

距今 6000 年前，古埃及已处于金石并用期。公元前 3000 年，上埃及国王美尼斯征服了下埃及，建立了统一的古埃及王朝。此后历经 31 个王朝，在公元前 4 世纪末，古埃及王朝被希腊马其顿王亚历山大征服。

由于气候原因，古埃及人很早就开始使用植物纤维。除了初期以棕榈纤维为主，亚麻一直是古埃及人的主要衣用布料。夏季的炎热与冬季的温和，使埃及人将亚麻布织得非常轻薄。又因为亚麻染色较难，故以白色为主，配着古埃及人红黑色皮肤，别具魅力（图 2–37）。

图 2–37　拉何泰普和内弗莱特坐像

绳衣腰衣

绳衣是古埃及最具特色和象征意义的服装之一。古埃及壁画上有许多只在腰臀部系一根细绳的年轻女子。从其环境和人物姿态推断，这些人应该是古埃及时期的奴隶或舞者（图 2–38）等，全身光素无饰，只在腰部系一绳。在古埃及早期，男子们一般上身赤膊，下身穿一块缠裹腰部的白色亚麻布或裹腰兜裆的腰衣“罗印 · 克罗斯”（Lion Cloth，图 2–39），与女子宽大长袍“卡拉西里斯”（Kalasiris）形成对比。这种绳衣与其说是服装，倒不如说是一种由社会文化所赋予的象征性装饰符号。

图 2–38　古埃及壁画中女性身穿的绳衣

早期的腰衣区别不大，但随着身份制度的显现，法老、贵族等上层阶级则以更多的布料浆洗成有褶裥装饰，外形挺括、棱角分明的三角形褶裙，甚至有的还饰有蓝色、黄色、绿色的纵向条纹。腰衣成为古埃及服装最富特色的性别元素。

图 2–39　古埃及男子腰衣形式

图 2-40　中王国时期的 Tunic 面料

图 2-41　身穿筒形紧身衣的古埃及女子浮雕

图 2-43　古埃及彩塑人像

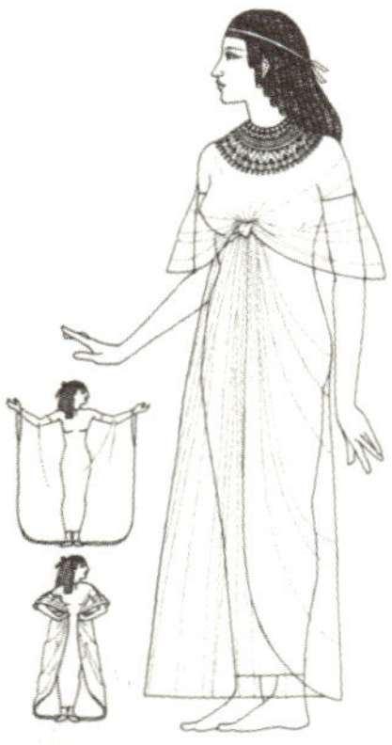
图 2-44　Kalasiris 着装示意图

图 2-42　公元前 3100 至公元前 2890 年第一王朝时期的 Tunic 实物

图 2-45　古埃及壁画中身穿 Lion Cloth 的男子和身穿 Kalasiris 的女子

丘尼卡（Tunic）

古埃及上层女性一般穿从胸到脚踝的筒形紧身衣丘尼卡（Tunic，图 2-40）。这是古埃及妇女的正式服装。古埃及女子的丘尼卡（Tunic）有筒裙或吊带连衣裙的形式，起自胸下直到踝骨，用背带吊在肩上或用腰带系住，乳房裸露在外，也有半袖连衣裙的形式。进入中王国时期，丘尼卡（Tunic）面料出现了多色手绘或纺织网状纹样（图 2-41）。

古埃及男性也穿丘尼卡（Tunic），但要比女性的丘尼卡（Tunic）短一些，腰身也更宽大。其式样如公元前 3100 至公元前 2890 年第一王朝时期的丘尼卡（Tunic）实物（图 2-42）。从图像资料上看，这种服装能将古埃及女性的身体轮廓毫无保留地显现出来，甚至在一些穿有丘尼卡（Tunic）的图像上还可清晰地辨别出腿部轮廓。因此，丘尼卡（Tunic）所使用的布料的伸缩性应该很好。这可能是借助百褶的形式得以实现。

卡拉西里斯（Kalasiris）

第十八王朝时期，好战的国王从美索不达米亚带回了许多战利品，其中就包括用一块布制成的宽大长袍卡拉西里斯（Kalasiris）。其裁剪方法是将一块相当于衣长两倍的麻布对折，在正中挖一个钻头的洞，两侧留出袖子部分后缝合（图 2-43 ～图 2-45）。细密褶裥所形成的丰富层次和明暗效果，使得埃及服饰极具魅力。服装也便于活动。其固定褶裥的方法是把衣料浸水、上浆、折叠、压紧后晾干。

与丘尼卡（Tunic）一样，卡拉西里斯（Kalasiris）是一种男女皆穿的服装。初期的卡拉西里斯（Kalasiris）非常宽松，长过脚踝，穿的时候在外面用带子系着，形成 X 型。到了第十八王朝末期，卡拉西里斯（Kalasiris）全身压成百褶，腰部的系带也有百褶，系扎后垂挂在腹前。

假发

自古以来，埃及人就留短发，进入王朝时代，无论男女都时兴戴假发（图 2–46），一方面是为了防晒，另一方面也与古埃及人的清洁癖有关。其形象如头戴假发身穿卡拉西里斯（Kalasiris）的古埃及女子木雕（图 2–47）。由于宗教仪式的规定，也为了清洁和防暑，古埃及男女皆剃发。男子剃秃，女子剪短，然后再戴假发，只有在服丧期间留头发和胡子。古埃及的假发是用网衬固定人发、羊毛或棕榈叶纤维制作成的（图 2–48）。由于在剃光的头上形成透气的空间，所以戴上这种假发很凉爽。根据习俗，古埃及假发的长短和社会地位成正比，国王和贵族们的假发最长。

图 2–47 头戴假发、身穿 Kalasiris 的古埃及女子木雕

图 2–46 第十九王朝法老宝座靠背上的浮雕图案

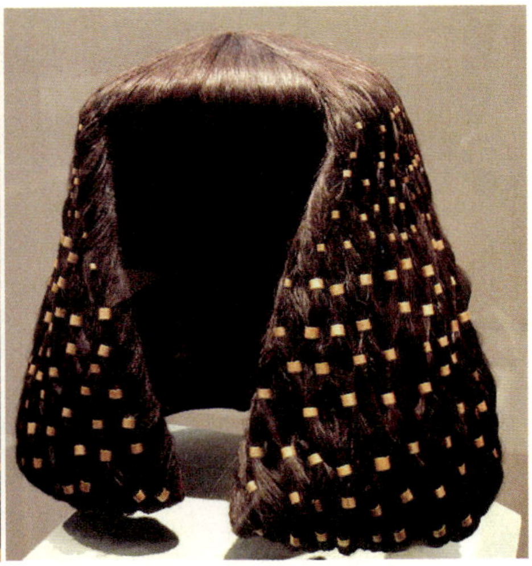
图 2–48 古埃及女子假发实物

图 2–49 古埃及女子眼部妆容

古埃及鞋尖有平底，也有上翘的造型，但最具特点的当属一种鞋尖细长，向上伸出反折，与第一和第二脚趾间及侧帮的系带固定的鞋。古埃及鞋匠还会用硬牛皮做凉鞋或短靴，用厚牛皮做鞋底。

图 2–51 古埃及遗址出土的黄金鞋

图 2–50 古埃及鞋子实物和线描图

化妆

女性很重视面部妆容（图 2–49），使用绿和黑墨画眉毛、眼圈和睫毛，用洋红色膏涂嘴唇，用白和红色粉涂脸颊，用橙色的散沫花汁涂染指甲。一般普通男性不留胡子，但帝王和高官例外，他们不但会留胡子，还在下巴处做成筒状。

鞋子

在埃及，最古老的鞋子是用植物制成的凉鞋（图 2–50），鞋底由草、藤或椰叶编成。军人作战时，打胜仗的一方就没收战败一方的鞋子作为战利品。古埃及鞋匠已经掌握了皮子染色技术，在古埃及遗址考古中，甚至有錾刻漂亮图案的黄金鞋出土（图 2–51），这应该是法老、贵族们等社会上层人物的用品。

二、美索不达米亚

（一）苏美尔人

美索不达米亚文明的苏美尔人，居住在两河汇流的下游苏美尔地区。他们是两河流域地区最早的居民。苏美尔人书写用的楔形文字，是人类使用最早的象形文字。

苏美尔人一般穿平整裹身并下垂到小腿处的圆形裙衣（图 2-52 ～图 2-54）。这类服装的另一重要特点是裙衣上的穗状垂片。考察雕像实物可知，这些服装的穗片有不同的款式，有的从膝上垂至膝下，有的则从腰间垂下，层层重叠成伞状裙卡吾纳凯斯（Kaunakes）。

后期的苏美尔人，一般上身裹一块搭过左肩的大围巾，头上戴着无边羊皮圆帽，它上面装饰着许多小波纹。苏美尔人服装面料以亚麻布与羊毛、羊皮为主。大约在公元前 8500 年，苏美尔人能获取的制衣所需纤维的动物是山羊和绵羊，植物则是亚麻。

苏美尔乌尔王陵出土的萨尔贡一世金盔头像反映了苏美尔人贵族束起的编发发式和有着优美波浪曲线的长须（图 2-55）。在乌尔城出土的乌尔军旗（the Standard of Ur）距今已有 4700 年，被公认为是两河流域最具代表性的文物。它是用贝壳、天青石与石灰石在木板上镶嵌出来的马赛克艺术品，有超过 100 个栩栩如生的人物与动物形象，很好地展示了当时苏美尔人的着装方式（图 2-56）。

图 2-52　拉伽什的饰板浮雕

图 2-53　穿 Kaunakes 的苏美尔人石雕人像

图 2-54　苏美尔人石雕人像

图 2-55　萨尔贡一世金盔头像

图 2-56　乌尔军旗

图 2-57 拉格什之王雕像

图 2-58 巴比伦女性着装形象

（二）古巴比伦

公元前 1750 至公元前 1595 年，统治美索不达米亚这一地区的是巴比伦人。巴比伦王国由塞姆族部落的阿卡德人建成。阿卡德人曾长期生活在苏美尔人的统治之下，故深受其影响。因此，巴比伦人的服装与苏美尔人的服装区别不大。

巴比伦男服的主要形式为缠裹式的袍服康迪斯（Candys），是用边缘有流苏装饰的三角形织物绕身包缠，形成参差不齐、错落相间、纵横繁复、雍容华贵的外观（图 2-57）。国王专用康迪斯（Candys）用金钱缝制，上面除了精致的刺绣图案外，还镶嵌珍贵的宝石。康迪斯（Candys）里面套有毛边流苏的直筒紧身长袍。紧身长袍有时单独穿着，其边缘有刺绣花饰，下摆亦有流苏饰物，造型简单，有时在外面披一块搭过单肩的披肩。在紧身衣上束有各种装饰讲究的腰带，有的腰带很宽，紧束腰身，更突出了男子宽阔的肩胸。

巴比伦女服与男服大体相同，只是略宽肥一些。巴比伦后期妇女的服装，是一种扣紧的披搭式服装，类似印度妇女所穿的莎丽。服装的面料是饰有流苏滚边的毛织物或亚麻，有红、绿、紫、蓝等鲜艳的颜色，有精美的刺绣装饰（图 2-58）。

巴比伦男子十分重视头发和胡须的卷烫整理，仔细地将胡须分段卷烫，显出层次和节奏，一层层卷曲的小波纹显得十分华丽而有层次。头发也同样经过卷烫整理，向后披着，并撒金粉。帝王和高官还配戴头冠，其上有一圈小的羽毛饰物，有时镶上宝石。当时的男女都戴着镶有珍珠及各种宝石的耳环、项链和手镯。巴比伦人还穿用皮革做成的凉鞋，款式和现在的凉鞋区别不大。鞋带在脚拇趾上绕一圈，然后绕在脚腕上并用扣子扣住。为了便于在林木中骑马奔跑和在激烈的战斗中保护双腿不受伤害，士兵们从脚尖到膝部都穿着用皮革制成的护腿。

图 2-59 亚述皇家长袍形象

图 2-60 浮雕上的亚述人形象

图 2-61 阿普利二世石雕及其披肩的穿搭方法

（三）亚述

公元前 14 世纪，亚述国兴起于底格里斯河中游，公元前 8 世纪到公元前 7 世纪势力扩大，征服了整个美索不达米亚地区，最盛期从现在的伊朗中部以西到土耳其东部，从叙利亚到西奈半岛，甚至包括埃及，形成了强大的帝国版图。

亚述人服装最大特征是大量使用流苏装饰。此外，刺绣和宝石装饰技巧也很发达，刺绣纹样的题材有埃及的莲花、棕榈、圣树、蔷薇和山梨等。亚述人服装所使用的面料主要是羊毛织物，同时，从埃及引进的亚麻，从印度引进的棉也逐渐得以普及，生产出了高质量的棉麻织物，有的还在织物中织进了金线。

从服装造型上看，当时外出服有两种基本样式：第一种，“丘尼卡（Tunic）+ 外套披肩式卷衣（披肩式）”，这是亚述男女装的共同特征（图 2-59～图 2-61）；第二种，“丘尼卡（Tunic）+ 外套贯头衣（斗篷式）”。披肩式服装肩部较宽，可达上臂中部。斗篷式服装由相同的两片布构成，前后两片在肩部缝合，留出领口，底摆为方形或圆形，衣服两侧不缝合。

图 2-62 雅典的帕特农神庙

三、爱琴文明

欧洲文明的源头在古希腊，古希腊文明的源头在西亚两河流域的美索不达米亚文明、希伯来文明和北非尼罗河流域的古埃及文明，以及地中海东部的爱琴文明。

爱琴文明是指公元前 20 世纪至公元前 12 世纪间的爱琴海域的上古文明，是希腊文明的源泉。古希腊文明是西方历史的开源，持续了约 650 年（公元前 800—公元前 146 年）。时至今日，我们仍可在西西里岛上一睹著名的神庙峡古、雅典的帕特农神庙（图 2-62）。

图 2-63 克里特岛诺萨斯宫殿壁画《戴百合花的王子》

图 2-64 克里特岛持蛇女神像

图 2-65 身穿短袖或长袖贯头式长袍的迈锡尼女性

图 2-66 迈锡尼武士群像

（一）米诺斯文明

克里特文明是爱琴海地区的古代文明，出现于古希腊、迈锡尼文明之前的青铜时代，主要集中在克里特岛。早自公元前 3000 年，克里特文明已进入铜器时代，出现象形文字，并有相当规模的建筑物。公元前 2000 年，克里特达到青铜时代全盛期，有宏伟的宫殿式建筑、各种精制的工艺品。

19 世纪末 20 世纪初，英国考古学家伊文发掘了克里特岛的诺萨斯宫殿遗址。从克里特岛发掘出土的壁画看，克里特男子的服饰较简单，上身一般赤膊，下身主要为包缠型短裙。克里特岛《戴百合花的王子》（图 2-63）壁画中的王子如真人般大小，头戴百合花和孔雀羽毛编成的帽子，长而略弯曲的长发顺肩而下，单剩一绺头发被王子在胸前随意轻拂着。他脖挂金色百合串成的项链，身着腰束皮带的短裙，腰身束得很细。

观察持蛇女神像（图 2-64），可以清楚地看到当时女装的式样：上身是一件近似现代立体裁剪服装的短袖紧身衣，前胸在乳房处敞开无扣，胸下则用绳扣紧紧束住，把乳房高高托起；腰部收得很细，下身为在臀部膨起的逐层扩展的喇叭裙，有六七层之多；整个衣裙的造型，从上到下把女性所具有的美丽曲线都衬托了出来；裙前还有一小块围裙状饰布；衣裙上的图案精美，色彩华丽。

（二）迈锡尼文明

迈锡尼文明是希腊青铜时代晚期文明，由伯罗奔尼撒半岛的迈锡尼城而得名。它继承和发展了克里特文明，是爱琴文明的重要组成部分。约公元前 2000 年，迈锡尼人开始在巴尔干半岛南端定居，到公元前 1600 年才成立王国。迈锡尼文明从公元前 1200 年开始呈现衰败之势，后来多利亚人南侵，宣告了迈锡尼文明的灭亡。

在迈锡尼的泰林斯城宫殿壁画遗迹中可以看到，当时女性服装有两种式样。第一种为短袖或长袖贯头式长袍，上面紧身贴体，裙摆较宽，呈吊钟形，在肩部和腋下侧缝处有条纹装饰（图 2-65）。此外，迈锡尼无论男女都留长发，并卷烫成一缕缕的发卷，飘逸在身前身后。

从考古实物分析，迈锡尼已经有了比较完善的武士盔甲。战车士兵穿的护胸甲是由皮衣上缝青铜片制成。防御性头盔是以绳系甲片连接而成，其下颌带起到稳固作用。他们使用木制或皮制的圆形和方形盾。进攻性武器大多是弓箭、青铜长枪、标枪和长剑（图 2-66）。

四、古希腊

古希腊服装的穿着方式包括裹缠和披挂式。其穿用最多的是希顿（Chiton）。这是一种男女同服的款式，女子希顿（Chiton）长至踝部，男子希顿（Chiton）长至膝部。因为民族和区域不同，希顿（Chiton）又分为多利安式希顿（Doric Chiton）和爱奥尼亚式希顿（Ioric Chiton）。

爱奥尼亚式希顿（Ioric Chiton）

爱奥尼亚式希顿（Ioric Chiton）是一种长至膝盖、两侧缝合的短袖束腰外衣（图 2–67、图 2–68）。穿的时候在肩袖部位用 12 枚别针固定，为了行动方便，用一根长长的系带将宽松的长衣随意系扎一下即可。古希腊人常用细绳把宽敞的“衣袖”捆扎在身上以便于劳作。其形象如德尔菲阿波罗神殿的青铜《驾驭者》（图 2–69）。

图 2–67 Ioric Chiton 穿着图

图 2–68 古希腊瓶画中身穿 Ioric Chiton 的女子形象

图 2–69 德尔菲阿波罗神殿的青铜《驾驭者》

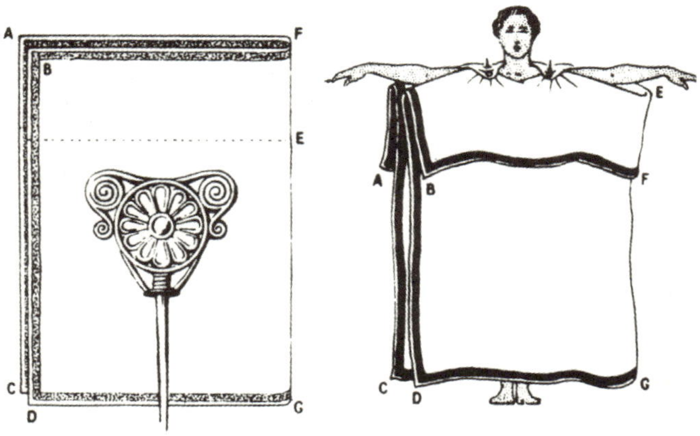

图 2–70 Doric Chiton 服装式样

图 2–71 雅典卫城帕特农神庙《哀伤的雅典娜》

多利安式希顿（Doric Chiton）

多利安式希顿（Doric Chiton）由一整块四方面料构成，长度超过穿者身高，宽度为人伸平两臂时右指尖到左指尖的两倍（图 2–70）。穿时先在上身处向外向下翻折一大块，翻折的长度随意，短可在腰线以上，长可至膝部；然后横向再做平均对折，包住躯干，两肩处用金属别针别住。对折的一边是敞开的，有时靠腰带固定，有时在腰际下或腋下缝死，成为宽肥的筒状裙。胸前和身后的翻折部分有时较长，可以系在腰带内，有时则自然悬垂或飘拂，形成细密垂褶，增加了平面衣料的立体感，充满明暗黑白不断转换的生动魅力。为了行动方便，古希腊人有时还用一根细绳系束在多利安式希顿腰部，形成堆积的褶皱（图 2–71）。

希玛申（Himation）

希玛申（Himation）是男女皆穿的包缠型长外衣，通常用长4～5米，宽1.2～1.5米的面料制成，最初为毛织物，穿时先搭在左臂左肩，从背后绕经右肩（或右腋下）再搭回到左肩背后（图2–72、图2–73）。衣料四角缀有小金属重物，以使衣角能自然下垂，外出时可以拉起衣服盖在头上防风雨，睡觉时可以脱下衣服当铺盖。缠绕型的服装主要依赖面料在人体上的围裹，形成延续不断、自由流动的褶裥线条，围裹的方式不同，所造成的款式各异。此外，古希腊女性偶尔会穿多层衣服，在多利安式希顿（Doric Chiton）外面套上爱奥尼亚式希顿（Ioric Chiton），甚至再在外面套上一件希玛申（Himation）。

图2–72　包裹头部的古希腊Himation

图2–73　古希腊壁画中的人物

迪普罗依斯（Diplois）

古希腊还流行一种叫迪普罗依斯（Diplois）的变形希玛申（Himation）。它也是用安全别针固定的服装（图2–74～图2–76），示意图中A部位用5枚安全别针固定后披在右肩，B部位垂挂在胸前，C部位垂挂在右臂的前后。

图2–74　古希腊大理石雕刻人像《雅典娜》

图2–75　身穿Diplois的古希腊雕塑人像

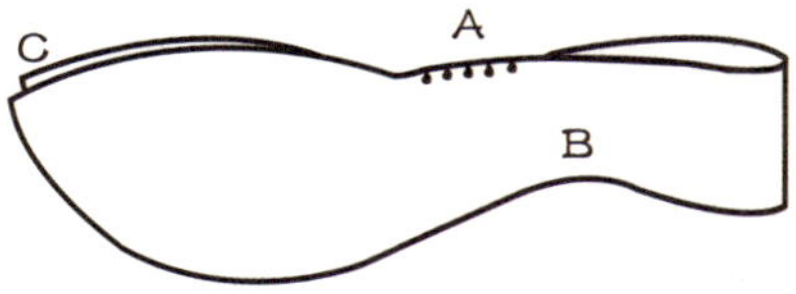

图2–76　Diplois示意图

图2–77　古希腊马赛克《猎狩雄鹿》

克拉米斯（Chlamys）

古希腊男子还穿一种叫克拉米斯（Chlamys）的小斗篷。外出旅行、骑马打仗时，古希腊男子一般将克拉米斯（Chlamys）披在左肩，在右肩用别针将面料两端扣住，露出右臂以便于活动（图2–77）。

第三章　公元前 2 世纪—公元 4 世纪

第一节　中国：汉代、魏晋南北朝

一、汉代

西汉：公元前 206—公元前 8 年
东汉：25—220 年

汉朝（公元前 206—公元 220 年）是继秦朝之后的大一统王朝，主要分为西汉、东汉时期。西汉初年，汉高祖刘邦加强中央集权，先后灭六国异姓王，稳定和巩固了中央政府的统治。随后，又经数十年的休养生息，在武帝时代，凭借文治思想，建全、完善了政府制度，奠定了坚实的经济基础和社会组织。

汉代是中国古代封建社会服饰制度的开端。永平二年（公元 59 年）正月，祀光武帝明堂位时，明帝及公卿诸侯首次穿着冠冕衣裳举行祭礼，成为中国封建社会以皇权地位为中心的儒家学说衣冠制度在中国得以全面贯彻执行的开端。

五时色

汉朝统治者认为汉承秦后，为土德，土是黄色，于是服色尚黄。方术家又把五行学说与占星术的五方观念融合（图 3–1），认为土象征中央；木是青色，象征东方；火是红色，象征南方；金是白色，象征西方；水是黑色，象征北方。青、红、黑、白、黄，为正色，为天子朝服颜色。又因天子统一天下，代表了天下各方的颜色，因而要求天子服装颜色须按季节不同而变换（图 3–2），即孟春穿青色，孟夏穿赤色，季夏穿黄色，孟秋穿白色，孟冬穿黑色。

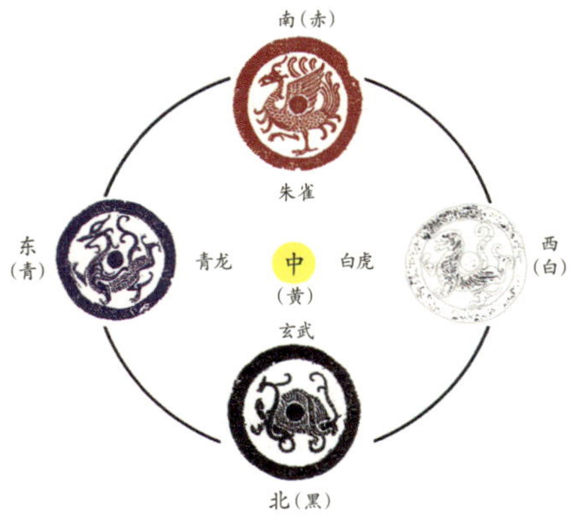

图 3–1　五行五色图

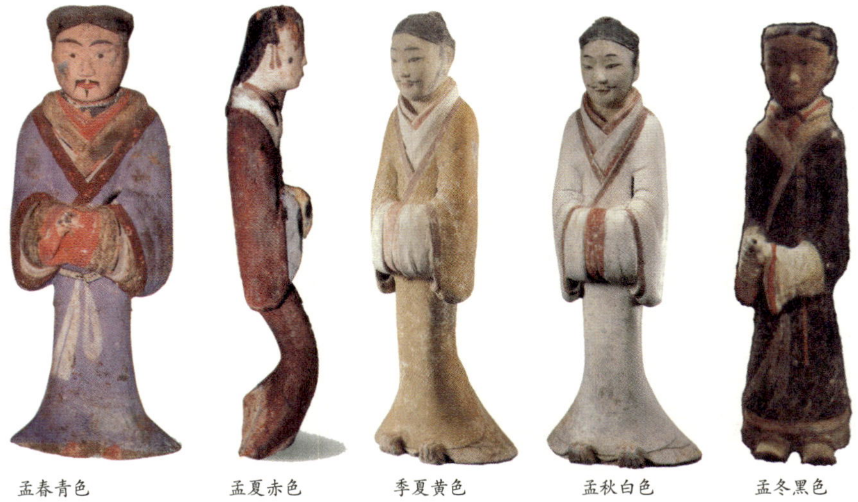

图 3–2　汉代彩绘陶俑

图 3-3　头戴长冠的彩绘木俑（湖南长沙马王堆西汉墓出土）

图 3-4　帛画女主人和侍者（湖南长沙马王堆西汉一号墓出土）

图 3-5　汉代汉阳陵彩绘陶俑

在一个四枚披垂的花叶基座中捧出一簇八根弯曲的细枝，除中间一茎立一只小鸟外，其余枝条顶端或结花朵，或结花蕾。

图 3-6　金步摇（甘肃武威西晋墓出土）

帝后服饰

据《后汉书·舆服制》规定，汉代皇帝服饰有祭服和常服两种。皇帝在祭祀天地明堂时，头戴冕冠，身穿有十二章纹装饰的玄色上衣和纁色下裳。当皇帝祀宗庙诸祀时，头戴长冠（图 3-3），外穿玄色绀缯深衣，绛缘领袖的中衣和绛色绔袜。当皇帝头戴通天冠时，身穿深衣形式的“五时色”袍服。

据《后汉书·舆服志》记载，汉代命妇朝、祭同服，即自二千石夫人以上至皇后的朝服都用助蚕时的服饰。其形制皆为深衣制的袍服，并在领、袖的部位以锦缘边，即所谓“隐领袖，缘以绦”。此时，衣襟绕襟层数又有所增加，下摆部分肥大，腰裹得很紧，衣襟角处缝一根绸带系在腰或臀部，如湖南长沙马王堆西汉一号墓出土帛画中的女主人服装形象（图 3-4）和汉代汉阳陵彩绘陶俑形象（图 3-5）。

汉代太皇太后、皇太后入庙穿绀（深青中有赤色）上皂下的深衣制袍服；助蚕穿青上缥下的袍服。翦牦帼，即以翦牦为廓匡，覆于假结之上。头上横簪，以安帼结。簪以檋瑁为擿，长一尺，其一端为华胜，上有以翡翠为毛羽的凤凰，凤凰嘴衔垂白珠，垂黄金镊左右各一个。此外，头上还有耳珰垂珠的珥。

皇后谒庙、助蚕的服饰与前者相同，步摇（图 3-6）簪珥，惟用假结，步摇以黄金为山题，贯白珠为桂枝相缪。一爵九华，熊、虎、赤罴、天禄（头上一角者）、辟邪（头上二角者）、南山丰大特（即南山丰水之大牛）六兽，即《周礼》中所谓的“副笄六珈”，诸爵兽皆以翡翠为毛羽，以黄金为山题，为白珠珰绕，以翡翠为华云。

冠帽

汉代时期，官员服饰上的等级差别标识主要为冠帽和佩绶。前者主要以职务区分为主，后者主要以官阶等级区分为主。

冠帽是汉代官服等级区分的主要标志。在汉代的首服制度中，有冕冠、长冠、委貌冠、皮弁冠、爵弁、建华冠、方山冠、巧士冠、通天冠、远游冠、高山冠、进贤冠、法冠、武冠、却非冠、却敌冠、樊哙冠、术士冠、鹖冠十九种之多，明代《三才图会》中有汉代部分冠帽式样。（图 3–7）。在汉代冠帽制度中，从冕冠至巧士冠属祭服类，从通天冠至术士冠属朝服类。汉朝政府根据不同的用途和等级分赐给官员，如文官戴进贤冠（图 3–8、图 3–9），武官戴武弁大冠，中外官、谒者、仆射戴高山冠，御史、廷尉戴法冠，宫殿门吏仆射戴却非冠，卫士戴却敌冠。因此，人们可以通过戴冠帽清楚地辨识其社会身份。戴不属于自己等级身份的冠无疑是严重违反礼规的行为。汉昭帝时，昌邑王刘贺让奴仆戴高山冠，被认为是“暴尊”的不祥行为。

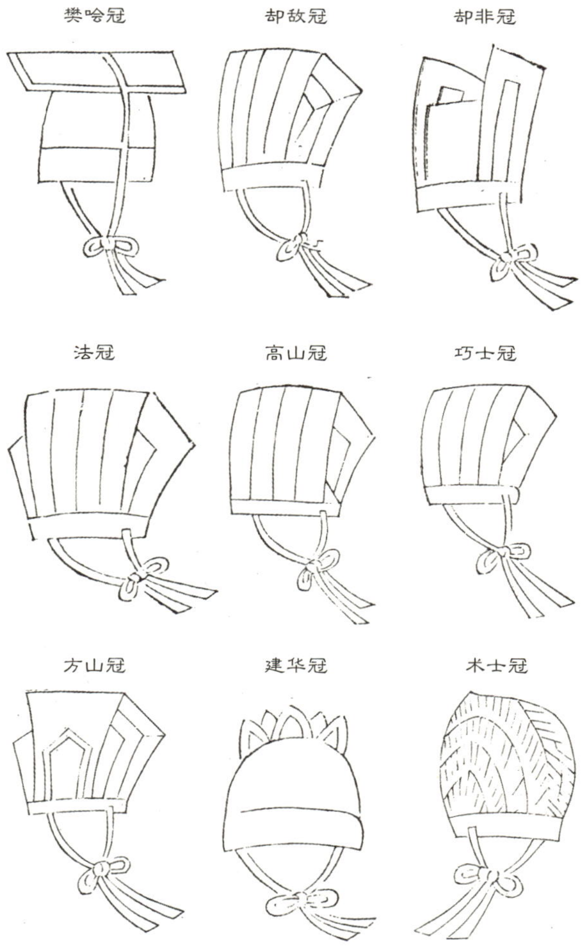

图 3–7　明代《三才图会》中汉代部分冠帽式样

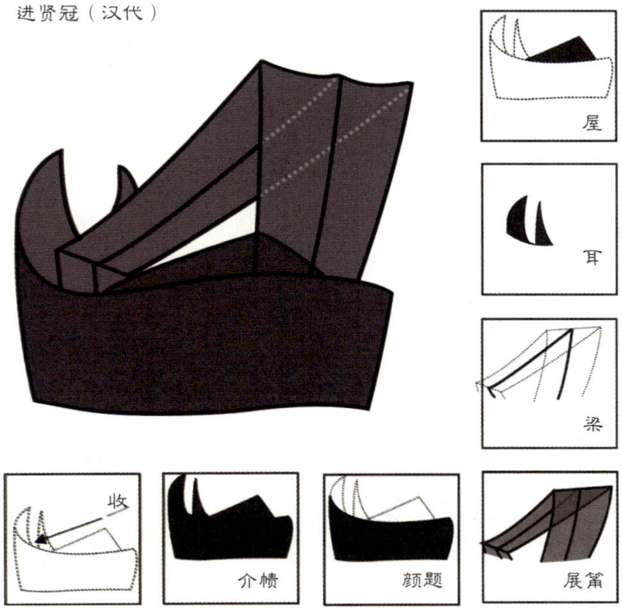

图 3–8　汉代进贤冠结构示意图（贾玺增绘制）

图 3–9　陶俑之进贤冠（湖南长沙金盆岭九号晋墓出土）

袍服

汉代服装主要有深衣、襜褕、袍、襌衣、褠、襦、衫等款式。其中的深衣、襜褕、袍、襌衣均属男女同服、上下分裁的一体式服装。它们在继承了自商周代以来形成的交领、右衽、缘饰、衿带等基本结构外，还具有多层穿衣，即“三重衣”的时代特征（图 3–10）。这三层袍服的彩色缘边分别在领口、袖口等处逐层显露，形成了丰富的层次对比。

东汉以前，袍服只是燕居时穿的一种日常便装。到了正式场合，汉人在袍服外要加罩表衣（深衣或襜褕）。到了东汉，袍服的地位逐渐上升，由内衣演变成外衣，并最终取代了襜褕，在一些隆重场合，如朝会、礼见时也可穿用。因为穿在外面，所以制作工艺讲究了许多。其领、袖、襟、裾等处多以丝绸缘边，即“衣作绣，锦为缘”，如湖南长沙马王堆西汉墓就有身穿曲裾缘边袍服的木俑出土。

为了御寒，袍服多有夹层。如果夹层是新丝绵称为“茧”，若是絮头或细碎枲麻等，则称为“缊”。汉代袍服的袖身多宽大，由袖身向袖口呈弧状逐渐上收。袖口窄小的边缘部分被称为“袪”；袖身宽的部分被称为“袂”，又因其形状与牛颈部相似，故又被称为“牛胡”（图 3–11）。牛胡本指牛颈部下垂的部分，引申指衣袖肘部宽博之处。其式样如湖南长沙马王堆西汉墓出土的直裾袍（图 3–12）和曲裾袍（图 3–13）。

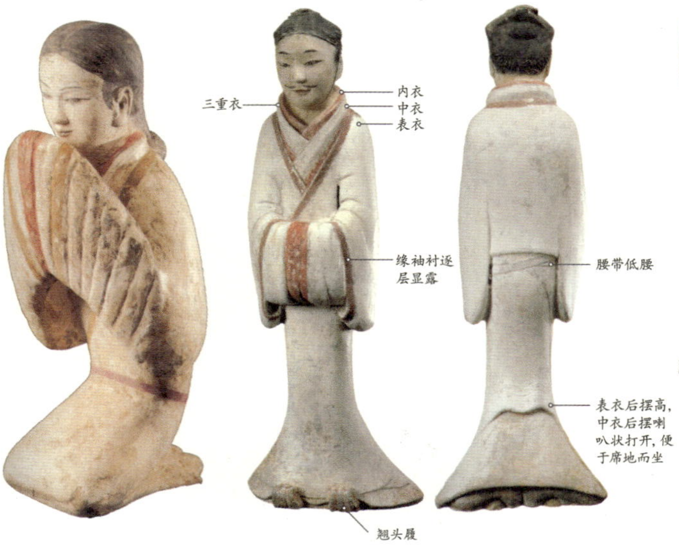

图 3–10　汉代人物陶俑（陕西汉阳陵墓出土）

图 3–12　“信期绣”褐罗绮绵直裾袍（湖南长沙马王堆西汉墓出土）

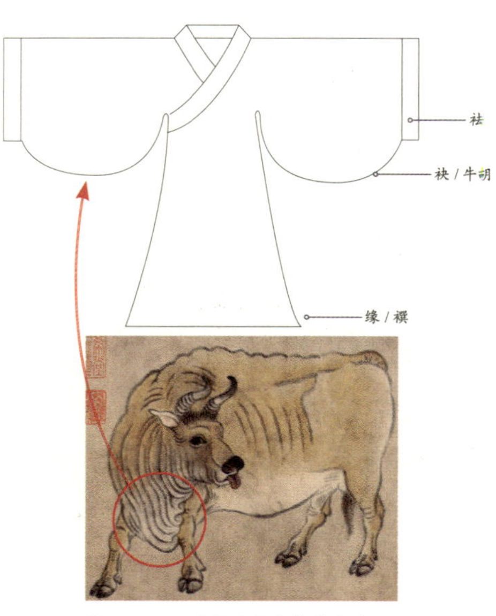

图 3–11　汉代袍服袖身部位名称

图 3–13　“信期绣”褐罗绮锦曲裾袍（湖南长沙马王堆西汉墓出土）

绶带

绶，又称“绂”和“组（组绶）”，是指以丝缕编成的带子。古代服饰中，绶既具有实用意义，又具有装饰功能。山东汉代画像砖《齐王赠绶钟离春》中的齐王手中托举的就是绶带（图 3-14）。从绶的起源来看，最初是作为连结玉器的实用物品，后来才逐渐演变成附属于祭服、朝服等礼服上的装饰品。

汉代政府对官员佩戴绶做了明确规定。佩戴者的地位越高，绶带越长，颜色越繁丽，织品的质地也越紧密。其实物应如尼雅发掘的组带（图 3-15）。

丑女钟离春是战国时代齐国无盐人氏，奇丑无比，年四十未嫁，但德才兼备，一心关心国家大事。齐宣王自恃其强，耽于酒色，齐国走向衰败。钟离春为了拯救国家，进谏齐宣王，陈述齐国危难四点，为宣王采纳，拆渐台、罢女乐、退谄谀，进直言，选兵马，实府库，齐国大安。后立钟离春为王后。

图 3-14　山东汉代画像砖中的《齐王赠绶钟离春》

图 3-15　1995 年新疆民丰尼雅遗址发掘的组带

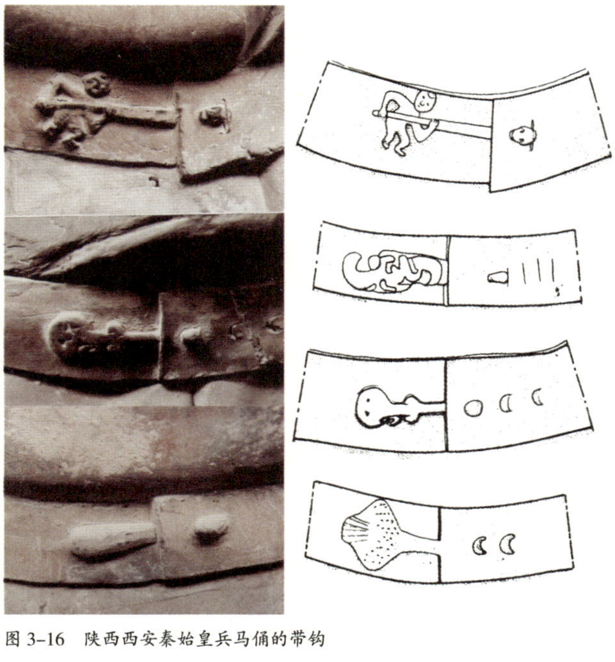

图 3-16　陕西西安秦始皇兵马俑的带钩

图 3-17　猿形银带钩（山东曲阜鲁国故城出土）

图 3-18　战国鎏金嵌玉镶琉璃银带钩

带钩

在秦汉以前，革带主要用于男装，妇女一般多系丝带，即《说文·革部》所载“男子革鞶，妇人带丝。”带钩是革带固定用的带头（图 3-16）。除装在革带的顶端用以束腰外，带钩也可用作衣襟的挂钩，还可装在腰侧用以佩刀、佩剑，钩挂镜囊、印章、刀剑、钱币等杂物。

中国古代带钩的造型可分为琵琶形、曲棒形、反勺形、耜形和鸟兽形五种。其中，猿形如山东曲阜鲁国故城出土的猿形银带钩（图 3-17）。后者通长 16.7 厘米，猿作振臂回首状，身微拱，目嵌蓝色料珠，通体贴金，背有一圆钮。工艺更复杂的是战国鎏金嵌玉镶琉璃银带钩（图 3-18）。

裙子

“裙”是“群”的同源派生词，意思是将多（群）幅布帛连缀到一起，成一个筒状，故刘熙《释名》释“裙”为“联结群幅也”。湖北江陵马山楚墓和湖南长沙马王堆西汉墓中出土的女裙正与此相互印证。湖南长沙马王堆西汉墓出土女裙两件（图 3–19），长 87 厘米，腰宽 145 厘米，下摆宽 193 厘米，接腰宽 3 厘米，均用宽一幅的绢四片缝制而成。考察陕西西安秦始皇兵马俑可知，男性也可将这种形制的短裙当作内衣穿着（图 3–20）。

图 3–19　女裙（湖南长沙马王堆西汉墓出土）

图 3–20　身穿裙式内衣的秦始皇兵马俑人俑

1972 年湖南长沙马王堆汉墓出土三件薄如蝉翼、轻若烟雾，仅重 49 克的单层长衣。出土遣册记为『素纱襌衣』。

图 3–21　素纱襌衣（湖南长沙马王堆西汉墓出土）

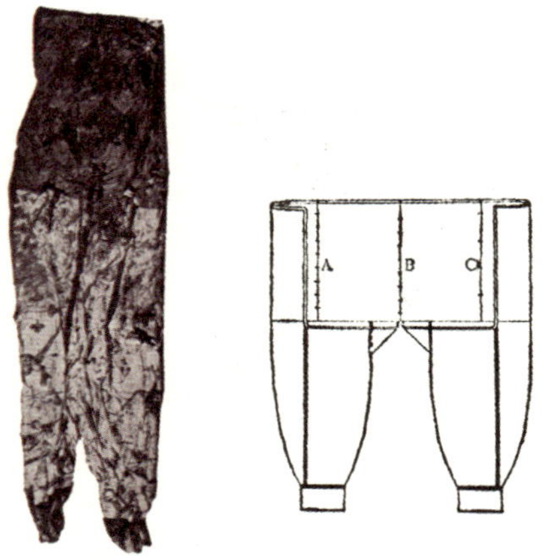

图 3–22　开裆绵裤（湖北江陵马山一号楚墓出土）

图 3–23　西汉空心砖上穿穷绔和短襦的武士形象

图 3–24　身穿合裆裤的汉代击鼓说唱俑

内衣

汉代有一种叫襌衣的内穿（图 3–21），刘熙《释名·释衣服》载：“襌衣，言无里也。有里曰复，无里曰襌。”

汉代人下身穿胫衣或开裆裤（时称“袴”，图 3–22）。由于无裆裤便溺方便，因此“开裆裤”又称为“溺绔”，故人们在公共场合的坐姿只能“跪坐”，两条腿分开伸坐和盘腿坐显然是不礼貌的行为。除了开裆，汉朝人有时也穿合裆裤（图 3–23、图 3–24）。穷绔一般和短襦配套穿用。为了御寒，有的裈内絮绵、麻等物，时称“复裈”。在中国古代服饰文化中，裤子外穿是一种有失礼仪的行为。西汉司马相如为了羞辱别人而只穿短裤出现在街市上。

犊鼻裈是贴身穿的短裤。山东嘉祥洪山汉墓、山东临沂沂南汉墓出土的画像石上就有穿着这种短裤耕作的农夫形象。这种短裤上宽下窄，两头有孔，以承受双股的贯串，与犊鼻之形十分相似，故称“犊鼻裈”。

长袖

楚国宫廷尚长袖舞，汉人继承楚人艺术，使得长袖舞更为盛行。舞女多是长袖细腰，有的腰身蜷曲，能使背后蜷成环状（图 3–25、图 3–26）。汉人傅毅《舞赋》载：“华袿飞髾，长袖交横，体若游龙，袖如素蜺。”汉刘邦妃子戚夫人的翘袖折腰之舞正是这种舞姿的体现。汉乐府《娇女诗》记载：“从容好赵舞，延袖象飞翮。”“延袖”就是指长袖。

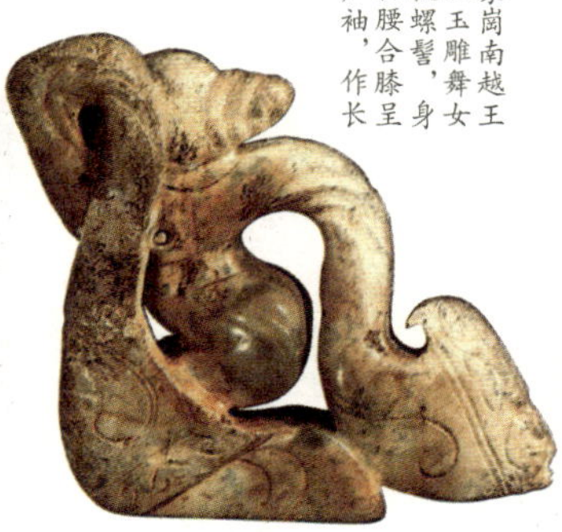

1983年广东象岗南越王墓出土的西汉玉雕舞女像，头右侧梳螺髻，舞女身穿长袖衣，扭腰合膝呈跪姿，舒展广袖，作长袖曼舞状。

图 3–25　西汉玉雕舞女像（广州象岗南越王墓出土）

玉雕舞女像是汉代特色佩饰，它在佩带时往往与其他玉饰一同组成串饰，挂佩于腰间或胸前。其外形飘飘欲仙，迎合了汉人对神仙飘游翱翔的想往。

图 3–26　汉代玉雕舞女像

湖南长沙马王堆西汉墓（共出土四双丝履）中出土青丝岐头履，长 26 厘米，头宽 7 厘米。该双层丝履，头部呈弧形凹陷，两端昂起分叉小尖角。

图 3–27　青丝岐头履（湖南长沙马王堆西汉墓出土）

面料主体花纹是神态不一的双脚兽，平列并排在忍冬藤蔓与花蕾变换的六列间隔之中，并填以小禽兽，还间有用绛色、白色、浅驼、浅橙和宝蓝等颜色织出“延年益寿长葆子孙”的隶书，袜口部分有一道金边，是织锦本身的缘边。

图 3–28　红罽绣花锦袜（新疆民丰汉墓出土）

图 3–29　东晋富且昌宜侯王天延命长织成履（新疆吐鲁番阿斯塔那墓出土）

履袜

汉代人如果穿着曳地长袍，脚下多穿鞋尖上翘的足服（图 3–27），上翘的鞋尖从前面将长袍的衣摆兜起，使行走方便。

袜是足衣，又写作“韈”“韤”。现存汉代时期的袜子实物，其质料多为罗、绢、麻及织锦等。汉代袜子讲究者以锦为之（图 3–28）。

魏晋时期，男女鞋履质料更加讲究，有“丝履”“锦履”“皮履”等。履的颜色也有一定制度，士卒百工服绿、青、白；奴婢侍服红、青。东晋织成履尤其精美，如晋人沈约诗云“锦履立花纹”。新疆吐鲁番阿斯塔那墓出土一双东晋富且昌宜侯王天延命长织成履（图 3–29）。其出土时色泽如新，用红、褐、白、黑、蓝、黄、绿等多彩丝线，以“通经断纬”方法在鞋面上织隶体汉字“富且昌宜侯王天延命长”，以及瑞兽纹、散花式小菱形花纹、倒山纹、忍冬纹。鞋底用麻线编织，耐用且轻巧。

革甲

据《周礼·冬官·考工记》记载，周代制甲业已取得一定经验，并设有“司甲”的官员掌管甲衣的生产。

周代的革甲由甲身、甲袖和甲裙组成。革甲分犀甲、兕（野牛）甲和合甲三种，质地非常坚硬。周代革甲形式如陕西西安秦始皇兵马俑身上所穿的甲胄（图 3-30 ～图 3-34）。除了士兵身上穿着铠甲，战马身上也有铠甲防护。

汉代戎服与秦代相似，军中不分尊卑，大都头戴平巾帻外罩武弁，身穿绵袍，腰束革带，下身套裤，足蹬圆头平底、月牙形头靴或履。1965 年，陕西咸阳渭城区杨家湾西汉大臣周勃、周亚夫父子的墓地出土的彩绘铠甲 2 000 余件，塑造了西汉皇家卫队的形象（图 3-35）。

图 3-30 秦始皇兵马俑跪射俑

图 3-31 秦始皇兵马俑站立俑背面

图 3-32 秦始皇兵马俑将军站立俑

图 3-33 美国《国家地理》杂志和美国华盛顿大学复原秦俑色彩

图 3-34 战国错金银刺虎镜上的骑士纹（传河南洛阳金村出土）

图 3-35 武士俑彩绘铠甲

二、魏晋南北朝

三国：220—265 年

西晋：265—317 年

十六国：304—439 年

东晋：317—420 年

南朝：420—589 年

北朝：386—581 年

图 3-36　传东晋顾恺之绘《洛神赋图》中的帝王形象

图 3-37　北朝进香图浮雕（美国费城大学博物馆藏）

图 3-38　江苏南京西善桥东晋墓拼镶砖画《竹林七贤与荣启期图》

魏晋南北朝 300 多年，战乱频繁，是中国古代历史上极为动荡分裂的时代。该时期先为魏、蜀、吴的三足鼎立，继而三国归晋（司马氏），在经过全国统一后，又分为西、东二晋，国内局势又趋纷乱。东晋之后南方是宋、齐、梁、陈。北方由北魏统一，后来分为西魏、东魏、北齐、北周。由此统称南北朝。在南北朝时期，尽管朝代众多，但每个朝代实际存在的时间都不长，仅二三十年左右。其中，惟北魏政权存在时间长达 148 年。

在这个特殊时期，中原地区政权的不稳定，为北方游牧民族进入中原地区带来了便利条件。战争和民族大迁徙，促进了胡汉杂居，南北交融，本土玄学、道教和外来佛教也趁势兴起。这使得汉族农耕文化与草原游牧文明和异域外来文化相互碰撞，形成了以农耕文明的礼仪服饰为主体，融合游牧文明的实用、便捷的日常服饰的形式。

魏晋风骨

魏晋南北朝时期的服饰演变也是各种思想相互碰撞、影响下的产物（图 3-36、图 3-37）。由于中原汉族王权的削弱，加之政治的腐败、“礼教”和经学的繁琐，使“独尊儒术”的信仰发生了动摇，玄学、道教及佛教禅宗思想盛行一时。受这些主张的影响，一大批游离于世俗，试图突破旧礼教的文人士大夫，除沉迷于饮酒、奏乐、吞丹、谈玄之外，也在服装上寻求发泄，以傲世荣。因此，“冠小而衣裳博大，风流相放，舆台成俗”成为风尚。江苏南京西善桥东晋墓拼镶砖画《竹林七贤与荣启期图》中刘伶、嵇康、阮籍等人着装就是此类“魏晋风骨”（图 3-38）。

除了小冠流行，魏晋时期，巾的地位起了显著的变化，逐渐成为士大夫阶层的常服。

大袖衫

魏晋南北朝时期，汉族服饰以单衣、对襟，大袖，袖无祛（收口缘边）的大袖衫为风尚。新疆营盘出土的北朝绞缬绢衣（图3–39）是目前中国能见到最早的一件大袖衫实物，具有极其重要的研究价值。《宋书·周郎传》载："凡一袖之大，足断为两，一裾之长，可分为二。"与这段文献对应的是河北磁县湾漳北朝墓出土的大袖衫人物陶俑（图3–40）。

图3–39 北朝绞缬绢衣（新疆营盘出土）

图3–40 身穿大袖衫人物陶俑（河北磁县湾漳北朝墓出土）

袴褶的裤是一种外穿的合裆散口裤，它与内穿的汉式胫衣、无裆袴和连裆裈的穿用场合区别迥然。南朝时的袴褶、衣袖和裤管更宽大，即广褶衣、大口裤。因为裤口肥大，为了更利于活动，人们在裤腿的膝盖下用丝带扎束，称为"缚袴"（图3–43）。

隋承北周旧俗，袴褶的使用更为广泛，不仅为皇帝田猎、皇太子从猎之服，也可做上下官员的公服、武官侍从之服和卤薄乐舞之服。到了唐代，袴褶成为更为正式场合穿的朝见之服。至唐末，袴褶之制始渐废弃。

袴褶

魏晋南北朝时，袴褶服成为最为常见的装束，除了用于家居闲处外，有时还用于礼见朝会。它是一种由上褶和下袴组成的二部式服装。

据《急就篇》记载："褶谓衣之最在上者也，其形若袍，短身而广袖，一曰左衽之袍也。"可知，褶是衣身较短，袖子宽大的上衣。褶始为西北少数民族的骑服，故初为左衽，在盛行于南北朝后，按汉族传统变为右衽。

图3–41 身着袴褶、外套裲裆的人物陶俑（河北磁县湾漳北朝墓出土）

图3–42 唐彩绘着裲裆甲文官俑（陕西礼泉郑仁泰墓出土）

图3–43 河南邓州彩色画像砖上的头戴卷檐帽、身穿襦衣、缚口裤的乐伎仪仗人物像

裲裆

与袴褶相配的还有裲裆。裲裆，亦作"两当""两裆"，其长仅至腰，无袖，只有前后襟以蔽胸背，《释名·释衣服》载："裲裆，其一当胸，其一当背，因以名之也。"裲裆最初是由两裆甲演变而来，在魏晋南北朝时期用作戎服和常服（图3–41、图3–42），至隋唐时期也用于朝服。裲裆既可保持躯干温度，还不增加衣袖厚度，男女都可穿着。

襦裙

魏晋时期，衣裳分离的“上襦下裙”“短襦长裙”的搭配形式开始流行。这与先前流行的连衣裳而纯之以采、长及足踝的深衣形成对比并同时流行。女服造型演变成“上俭下丰”的式样（图 3–44）。曹植《洛神赋》载：“秾纤得衷，修短合度。肩若削成，腰如约素。”南梁庾肩吾《南苑看人还诗》云：“细腰宜窄衣，长钗巧挟鬟。”这都是在歌咏魏晋时期女性的窄式衣装之美（图 3–45、图 3–46）。

上下一体式女袍的腰线已经由胯部提升至腋下，用腰带系束，袍摆很大，拖地成椭圆状。

图 3–44　东晋顾恺之绘《女史箴图》

图 3–45　东晋女陶俑（江苏南京出土）

上下分体式的襦裙女装也是上衣合体窄小，裙多褶裥肥大，裙摆及地。

图 3–46　长裙妇女彩绘俑（陕西西安草场坡出土）

图 3–47　西汉彩绘骑兵俑（陕西渭城杨家湾墓地出土）

图 3–48　彩绘木俑（湖北江陵凤凰山 168 号墓出土）

图 3–49　彩绘《列女古贤图》漆屏中《周室三母》（山西大同北魏司马金龙墓出土）

袿衣

袿衣是一种以缯为缘饰，曲裾绕襟，底部有裾，形成的上宽下窄、呈刀圭形的两尖角的长襦（图 3–47）。汉代刘熙《释名·释衣服》载：“妇人上服曰袿，其下垂者，上广下狭，如刀圭也。”女子穿袿衣时，里面一般穿裙子（图 3–48）。

飞髾

魏晋流行缀以尖角（时称“襳”）和飘带（时称“髾”）的服饰。这些飘带走动时显现出天衣飞扬和乘风登仙的气韵，故有“华带飞髾”“扬轻袿之猗靡”的描写（图 3–49）。此时，女性还在盘成的双髻和高髻后垂下一撮头发，时称“垂髾”或“分髾”。

第二节　西方：罗马

传说在公元前 753 年，被古希腊人打败的特洛亚人的后裔，从小亚细亚渡海来到意大利，在台伯河边建立了罗马城，后来逐渐强大并征服了整个意大利。公元前 1 世纪，它征服了古希腊，称霸地中海，成为横跨欧亚非三洲的奴隶制大帝国。为了便于统治，4 世纪末，出现了两个皇帝东西分治的情形。东罗马以君士坦丁堡（即土耳其的伊斯坦布尔）为首都，又称拜占庭帝国。476 年，西罗马被日耳曼人所灭后，失去本土的东罗马帝国开始向封建社会过渡。

古罗马文化和古希腊文化有着密切联系。当古罗马人征服古希腊以后，更是对古希腊的文化艺术大加推崇和发扬，与古罗马文化融会贯通。古罗马服饰的面料有轻软的羊毛织物和亚麻布，后期的罗马还从东方引进了昂贵的轻薄美丽的丝织物。

一、托加（Toga）

托加（Toga，也称罗马长袍）是一种象征罗马人身份的披缠外衣。在王政时代的 250 年间男女均可穿着，到共和制时代，只有男子才能穿着，女子只能穿斯托拉（Stola）及帕拉（Palla），没有罗马公民权者则被禁止穿着托加（Toga）。它和希腊裹缠型外衣希玛申（Himation）近似，不过希玛申（Himation）为方形，托加（Toga）则呈半圆形（图 3–50）。

托加（Toga）最显著的特点是其超大、超长的尺寸——长约 540 厘米，最多达穿者身高的三倍；宽约 180 厘米，最多达穿者身高的两倍（图 3–51）。其半圆的两尖端垂有小重物，其中一端在前身膝部下垂，上面从胸前向左肩披过，经背后绕至右腕或右腋下，经过胸前再搭回左肩并在后背垂下，整个右腕或左臂可以自由活动。

古罗马女性穿托加（Toga）时往往会在里面套从古希腊女装那继承下来的希顿（Chiton），并把头部也一起裹缠起来（图 3–52）。

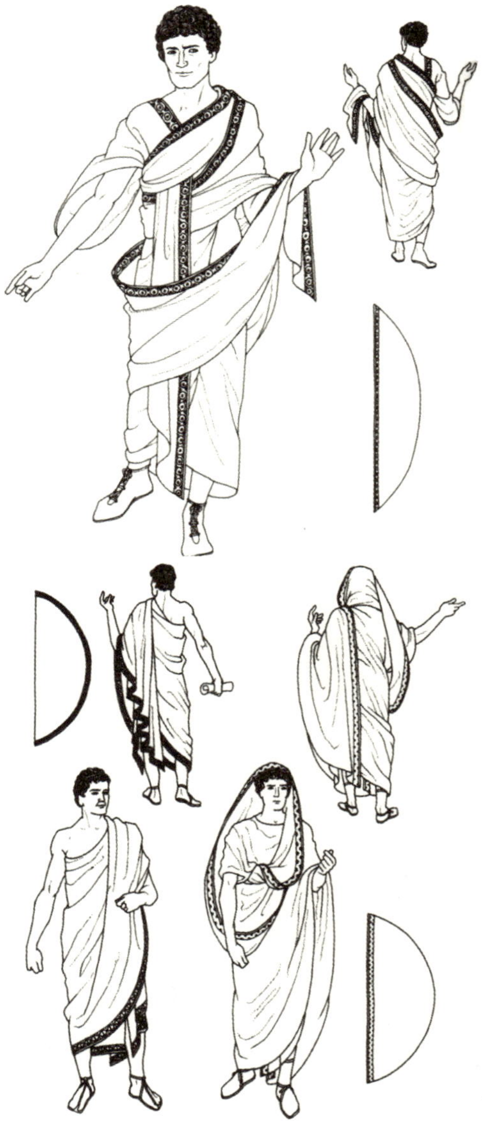

图 3–50　古罗马的 Toga

图 3–51　提着两个祖先头像的贵族

图 3–52　身穿 Toga 的罗马女性

丘尼卡（Tunic）

丘尼卡（Tunic）是一种有连袖的宽敞筒形衣（图 3-53），有时可穿内外两件，内层较短小，外层较长大，作为平日的简便装束，必要时会将托加（Toga）套在外面。古希腊人一般只穿单层衣服，而古罗马人习惯穿双层衣服。因此，内层衣就逐渐变为紧身的长筒形衬衫，一般长于膝下，男服用羊毛织物制作，女服用亚麻布制作。

佩奴拉（Paenula）

佩奴拉（Paenula）是一种钟形的可防寒、防雨的外套。多数佩奴拉（Paenula）是在整片面料上开领洞制成，但偶尔也有在身前留有开口的式样（图 3-54）。

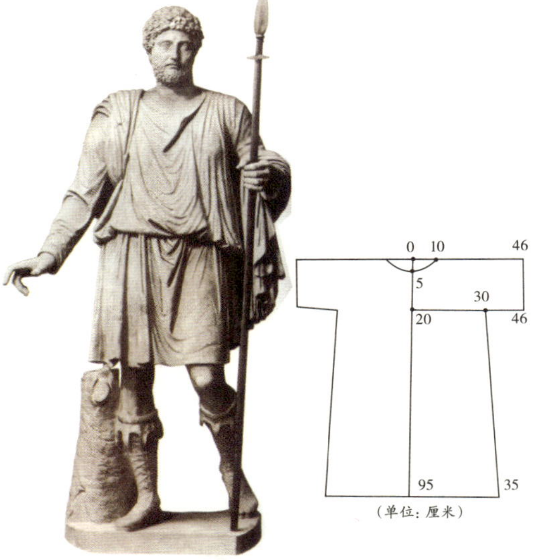

图 3-53　身穿 Tunic 的古罗马人雕塑及裁剪尺寸图

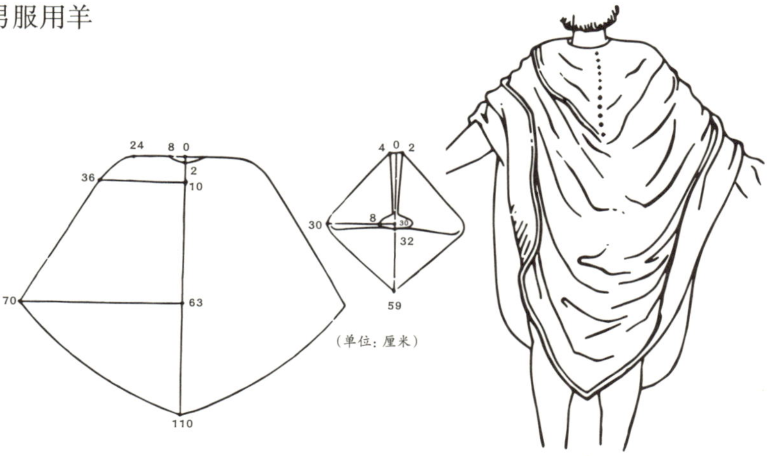

图 3-54　Tarquinia 地区 Triclinium 墓壁画中身穿 Paenula 的人物像

图 3-55　古罗马戎服

图 3-56　古罗马鱼鳞甲和板甲

图 3-57　普莱马波尔塔的奥古斯都

盔甲

讲究奢侈而又好战的罗马人，在军服设计上也用尽了心思（图 3-55）。古罗马盔甲有鱼鳞甲、锁子甲，以及源自古希腊并经过优化设计的板甲（图 3-56）。士兵们在盔甲外面套红色或白色长袍，高级指挥官则披白色羊毛斗篷。最高统帅即皇帝的铠甲非常华丽，上面有精致而优美的浮雕（图 3-57），铠甲下面有短的百褶裙。好莱坞电视剧《罗马》中再现了凯撒大帝与罗马兵团（图 3-58）。

图 3-58　好莱坞电视剧《罗马》中的凯撒大帝与罗马兵团

图 3-59 古罗马壁画中女子形象

图 3-60 身穿 Stola 和 Palla 的古罗马女性雕像

图 3-61 古罗马壁画中的女子形象

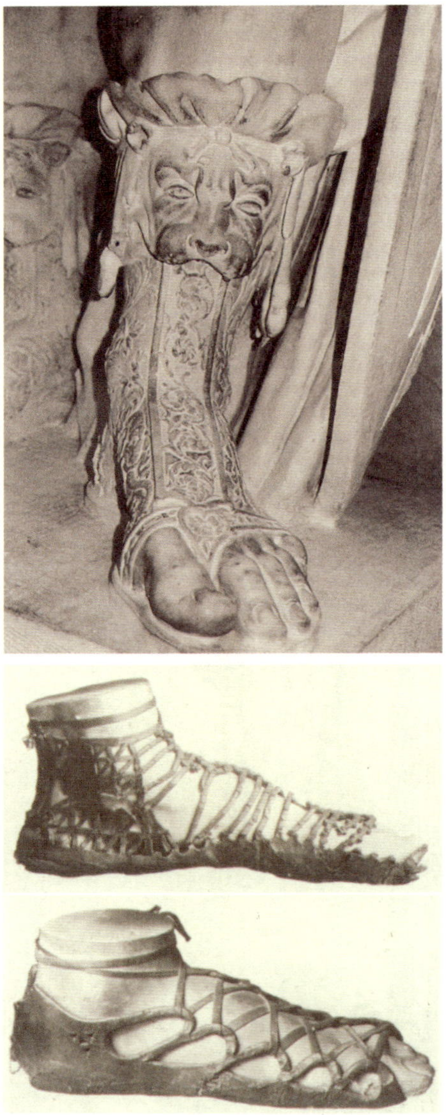

图 3-62 古罗马武士靴和武士鞋

帕拉（Palla）

女式披肩帕拉（Palla）是古罗马女性缠绕在丘尼卡（Tunic）或斯托拉（Stola）之外的一块大约 9 英尺（2.74 米左右）长、5 英尺（1.5 米左右）宽的长方形麻或毛织物裹缠衣（图 3-59、图 3-60）。

有时，帕拉（Palla）还被拉到头上兼作面纱。当妇女坐着或者处于一种更加放松的姿势时，也可以把帕拉（Palla）披挂在手臂上。妇女外出时，要用帕拉（Palla）遮盖住自己的身体和头部（图 3-61）。在古罗马人看来，这样做既可以避寒或者避免不得体的显露，又可以避开任何邪恶的眼睛。

二、罗马鞋

古罗马人在制作靴鞋方面，效仿和继承了古希腊人的高超技艺。平时，罗马人在室内穿着用皮革或草席制成的拖鞋。上街穿皮鞋，鞋的脚跟和脚面部分用大块皮子做出合脚形状并用带子系结，脚趾露在外面。有一些露趾武士靴，上面装饰着精美的花纹和作凶悍状的兽头，极其精美奢华。古罗马时期带有粗犷风格的角斗士鞋（Gladiator Sandals）为平底，以漆皮或彩色皮质拼接，并带有精致的交叉绑带设计，沙土色、咖啡色、原皮色等是古罗马角斗士鞋（Gladiator Sandals）的主流色彩（图 3-62）。

第四章　公元5—10世纪

第一节　中国：唐代

隋：581—618年
唐：618—907年
五代：907—960年

618年，李渊建立大唐王朝。唐朝国力强盛，军事强大，军事力量威震欧亚。京城长安不仅是国内政治、经济、文化中心，也是当时颇具影响力的国际性大都市。当时与唐朝政府有过往来的国家和地区曾达300多个，正如诗人王维《和贾至舍人早朝大明宫之作》所描绘的“万国衣冠拜冕旒”的盛况，每年有大批的留学生、外交使节、客商、僧人、歌舞艺人往来于长安。

图4-1　唐贞观时期壁画《维摩诘说法图》中身穿冕服的唐代帝王像（敦煌莫高窟220窟）

一、汉式礼服

隋唐时期，中国在政治上又一次南北统一，但服装却分成两类：一类继承了中原地区农耕文明传统的汉式冠冕衣裳，用作祭服（图4-1）、朝服（图4-2）和较朝服简化的公服；另一类则吸取了北方游牧民族的特点，使用便捷、实用的幞头、圆领缺骻袍和乌皮靴，用作平日的常服。隋唐以前，中国历代政府所制定的各种服饰制度多针对朝服和祭服。隋唐以后，公服和常服也纳入了服饰制度的范围，从而补充和完善了中国古代服饰制度的建设。农耕文明和游牧民族服饰的双轨制形成，适应了中国古代社会礼仪制度和日常生活的实际需要。

图4-2　唐景云二年《礼宾图》（陕西乾县乾陵陪葬墓张怀太子李贤墓墓道东壁壁画）

品色服

隋唐开始，标志官职身份的汉式冠帽被大力简化，无官职等级区分的常服使用普遍，促成了以不同服色区分身份贵贱、官位高低等序的“品色服”制度。唐上元元年（674年），唐政府规定“武官三品以上服紫，金玉带；四品深绯（图4–3），五品浅绯，并金带；六品深绿（图4–4），七品浅绿，并银带；八品深青，九品浅青。”九品职官服色各异，清楚地显示出穿者的身份等级、品序大小（详见表4–1）。

表 4–1　唐代品官服色制度

品色	紫	深绯	浅绯	深绿	浅绿	深青	浅青
官级	三品以上	四品	五品	六品	七品	八品	九品

图 4–3　彩绘武官俑（甘肃庆城赵子沟穆泰墓出土）

图 4–4　唐三彩文官俑（陕西咸阳礼泉出土）

唐代大诗人白居易《琵琶行》载：“座中泣下谁最多？江州司马青衫湿。”其表露了这种由政治地位变化而反映于服色变化的情况，以及由此带来诗人黯然无奈的心境。比青衫略高一级的是浅绿，元稹《寄刘颇二首》载：“无限公卿因战得，与君依旧绿衫行。”浅绿如甘肃庆城县赵子沟村穆泰墓出土的彩绘武官俑所穿上衣颜色。比浅绿等级更高级的是绯色和紫色，如元稹《酬乐天喜邻郡》载：“蹇驴瘦马尘中伴，紫绶朱衣梦里身。”白居易《郡中春宴，因赠诸客》载：“暗澹绯衫故，斓斑白发新。”

图 4–5　传阎立本绘《步辇图》中的圆领袍

自从“品色服”制度实施后，“青红皂白”便成为了官服和民服用色的界限分野。

圆领常服

唐代常服为幞头、圆领袍、銙带、乌皮靴。

圆领袍（图 4–5），亦称盘领袍、团领袍，是指右襟需固定于左肩之上，然后顺左肩向腋下掩盖。其领部正适合纽扣闭合，而圆领袍属于胡服，是隋唐时期士庶、官宦男子在祭祀、典礼等礼仪场合之外的日常服装。因为圆领袍在唐代属于常服，所以各阶层都可穿着。新疆吐鲁番阿斯塔那古墓群西州张雄夫妇墓出土木俑身穿的也是黄绢圆领袍（图 4–6）。

图 4–6　圆领袍木俑（新疆吐鲁番阿斯塔那古墓群 206 号墓西州张雄夫妇墓出土）

幞头

幞头是唐代男子常服的首服。甘肃庆城赵子沟穆泰墓出土的身穿翻领锦袍的彩绘陶胡人俑头上表现的就是幅巾覆头前后系结的幞头（图 4–7）。其内还有扣覆在发髻上的巾子（图 4–8）。巾子是以竹篾、藤皮及芒草等材料编织而成。新疆吐鲁番阿斯塔那唐墓出土的两件巾子实物均以丝葛类织物浸漆模压而成，顶部分瓣明显，表面布满菱格，左右两侧留有插笄的圆孔。其系裹方法是先使用布帕包头，前面的两角绕至颅后缚结，结带飘垂，后两角朝前包抄，系结于额，曲折附顶（图 4–9）。

图 4–7　身穿翻领锦袍的彩绘陶胡人俑（甘肃庆城赵子沟村穆泰墓出土）

图 4–8　巾子实物（新疆吐鲁番阿斯塔那唐墓出土）

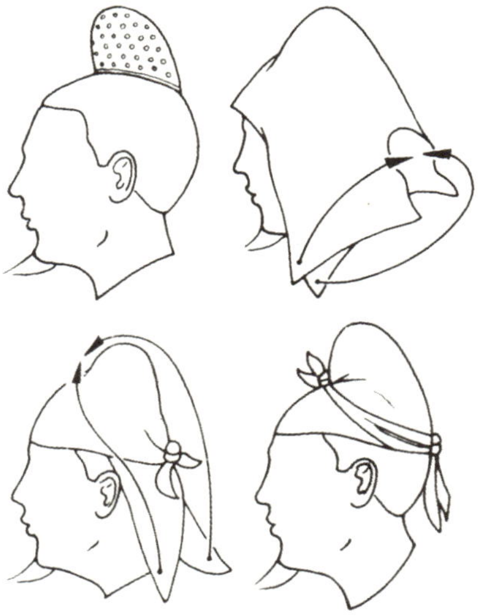

图 4–9　唐代幞头系扎示意图

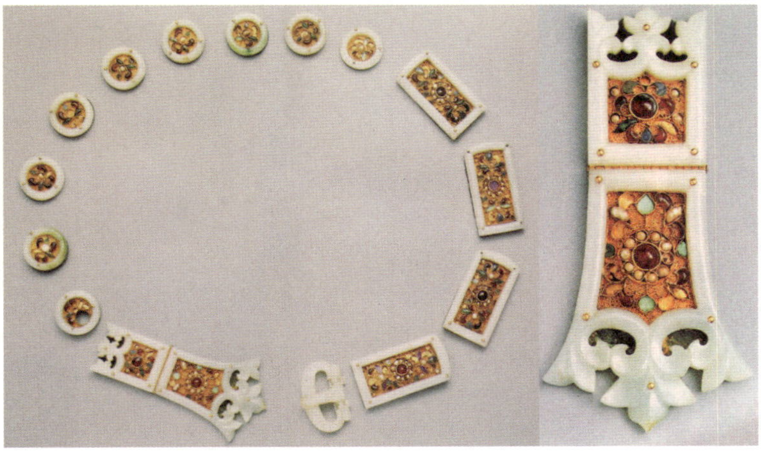

图 4–10　唐镶金嵌珠宝玉带饰（陕西西安南里王村窦皦墓出土）

图 4–11　明人绘《岳飞画像》

銙带

銙带由带鞓、带銙、带头、带尾四部分组成。带鞓是带的主体，由皮革做成，外裹红色或黑色绢。带銙是镶在带鞓上的牌饰（图 4–10），是腰带等级区别的标志，《新唐书·舆服志》载：“一品、二品銙以金，六品以上以犀，九品以上以银，庶人以铁。”带头，即带扣。带尾，也称“铊尾”，腰带系束之后，带尾朝下，表示对朝廷的顺服。

宋代銙带等级更为细致繁复（图 4–11）。同为金，有花纹不同者；花纹同者，有重量区别。宋朝公服的腰带、牌饰允许用玉，品官可用犀角，民、郡县吏、伎等人的腰带可用铜、铁、角、石。牌饰图案主要有毬路、御仙花、荔枝、师蛮、海捷、宝藏、天王、八仙、犀牛、宝瓶、双鹿、行虎、窪面、戏童、胡荽、凤子、宝相花和野马。

袒胸

中国传统封建礼教对女性的要求严格，不仅约束其举止、桎梏其思想，还要将其身体紧紧包裹起来，不许稍有裸露。但唐代国风开放，女子的社会地位得到极大提高，生活空间也更为广阔，着装风气也开先河，以袒颈露胸为时尚。袒胸装的流行与当时女性以身材丰腴健硕为佳，以皮肤白晰粉嫩、晶莹剔透为美的社会审美风气是分不开的（图 4–12）。李洞《赠庞炼师》载："两脸酒醺红杏妒，半胸酥嫩白云饶。"李群玉《同郑相并歌姬小饮戏赠》载："胸前瑞雪灯斜照，眼底桃花酒半醺。"这些唐诗均是对这种风尚的描写。最初，袒胸装多在歌伎舞女中流行，后来宫中和社会上层妇女也引以为尚，纷纷效仿。传唐代周昉绘《簪花仕女图》中的仕女个个体态丰满，粉胸半掩，极具富贵之态（图 4–13）。学术界对辽宁省博物馆馆藏《簪花仕女图》的创作年代看法不一，杨仁恺持中唐说、徐邦达持中晚唐说、谢稚柳持南唐说、沈从文则持宋代说）。从唐代中后期开始，随着胡服影响渐弱，袒胸装也逐渐消失。

图 4–12 唐三彩釉陶女立俑（陕西西安西郊纺织厂出土）

图 4–13 传唐代周昉绘《簪花仕女图》

罗衫

唐代贵妇上衣一般是对襟的窄袖衫或大袖衫。前者属于日常类便服，穿时衣襟敞开，下束于裙内。衣袖短者至腕，长者掩手。唐人绘《挥扇仕女图》（图 4–14）、《内人双陆图》和《调琴啜茗图》中均绘有着窄袖衫的妇女形象。

图 4–14 唐人绘《挥扇仕女图》局部

大袖衫属女子礼服类，流行用纱罗等轻薄材料制成，衣长至胯以下。唐代诗词中有很多描写女衫轻薄的优美诗句，如张祜《感王将军柘枝妓殁》载："鸳鸯钿带抛何处，孔雀罗衫付阿谁。"唐代女子穿着大袖衫时，衣摆多披垂于裙身之外。

唐末五代，薄衫金缕成为风尚，如裴虔余《柳枝词·咏篙水溅妓衣》载："半额微黄金缕衣，玉搔头袅凤双飞。"实物如 1981 年陕西宝鸡法门寺地宫出土的红罗地蹙金绣随捧真身菩萨佛衣模型。其中的红罗地蹙金绣半臂衫用绢做里，用降色罗做面，上面均匀分布蹙金绣折枝花卉纹样。花朵外面衬花叶，每朵花都留出一颗花蕊（图 4–15）。

图 4–15 红罗地蹙金绣半臂衫（陕西宝鸡法门寺地宫出土）

石榴裙

唐代女服以短襦长裙（图4–16）为尚。孟浩然《春情》诗云："坐时衣带萦纤草，行即裙裾扫落梅。"唐文宗时，曾下令禁止裙摆过长。唐代裙色以红裙为尚（图4–17），时称"石榴裙"（石榴花染色）、"茜裙"（茜草浸染），元稹《樱桃花》诗云："窣破罗裙红似火"。

间色裙

唐代还流行两色布帛相拼的"间色裙"。《旧唐书·高宗本记》载："其异色绫锦，并花间裙衣等，糜费既广，俱害女工。天后，我之匹敌，常着七破间裙。""七破"指裙上被剖成七道，以间它色，拼缝而成。奢侈者更可达"十二破"之多（图 4–18、图 4–19）。

图 4–16 唐代舞女绢画（新疆吐鲁番阿斯塔那墓张礼臣墓出土）

图 4–17 昭陵唐墓壁画中身穿红色长裙的唐代女性形象

图 4–18 昭陵唐墓壁画身穿间色裙的唐代女性形象

图 4–19 唐代木身锦衣裙仕女俑（新疆吐鲁番阿斯塔那张雄夫妇墓出土）

图 4–20 唐代宝相花夹缬褶绢裙（新疆吐鲁番阿斯塔那墓出土）

翠裙

绿色之裙也深受唐代妇女青睐，时有"碧裙""翠裙""翡翠裙"之称（图 4–20）。其实物如新疆吐鲁番阿斯塔那墓出土的唐代宝相花夹缬褶绢裙与陕西西安王家坟出土的唐三彩女乐俑下身所穿的高腰十字瑞花条纹绿色锦裙（图 4–21、图 4–22）。

图 4–21 唐三彩女乐俑（陕西西安王家坟出土）

图 4–22 唐三彩女俑（故宫博物院藏）

女帽

唐朝政策开放，经济繁荣，文化昌盛，对外交往频繁，兼收并蓄异域文化。唐代女性的社会生活也呈现出一种开放态势。在开元天宝年间（713—741年），西域文化大规模传入，妇女们或女效男装，着圆领袍，头裹幞头；或学胡服，多穿翻领窄袖袍，头戴胡帽。

唐中期，汉胡文化大融合，胡舞盛行。人们从对胡舞的崇尚，发展到对胡服的模仿，由此出现了元稹《和李校书新题乐府十二首·法曲》一诗中“女为胡妇学胡妆”的现象。唐代女性穿的胡服集中表现在羃䍦、帷帽、胡帽和回鹘装的流行上。

隋朝西域地区有人穿戴用轻薄纱罗制成的大幅方巾，用以遮蔽路上的扬尘。唐初武德至贞观年间（618—649 年），中原女性已开始有戴羃䍦的风气了。这正符合中国传统礼节中要求女性“出门掩面”的封建礼俗。羃䍦在日本东京国立博物馆收藏的一幅唐人绘画《树下人物图》中有所反映（图 4–23）。与此类似的人物形象还见于 1972 年新疆吐鲁番阿斯塔那 187 号墓出土的唐三彩仕女俑（图 4–24）。

永徽年间（650—655 年），随着社会风气的开放，人们开始使用一种“拖裙到颈，渐为浅露”的帷帽，这是一种在席帽帽檐周围加缀一层面纱的改良首服。唐高宗认为这样有伤风化，于是下敕禁止。但至天宝年间（742—756 年），唐代妇女头戴帷帽已成为一时风尚，如新疆吐鲁番阿斯塔那 187 号墓出土的头戴帷帽的彩绘陶女俑（图 4–25）。

胡帽，又称“蕃帽”，主要是指唐代及之前由西北或北方传入并在中原地域流行的皮帽或毡帽。与羃䍦和帷帽相比，胡帽在中原地区流行的时间相对较晚。其特点是帽子顶部尖而中空，珠帽、绣帽、搭耳帽、浑脱帽、卷檐虚帽等都可归为胡帽（图 4–26 ～图 4–28）。比起“全身障蔽”的羃䍦和将面部“浅露”于外的帷帽，它更加“解放”了，使得女性“靓妆露面，无复障蔽”。从羃䍦到帷帽，再到胡帽的发展，是妇女服饰史的进步与发展，反映了唐代社会开放的风尚。

图 4–23 唐人绘《树下人物图》（日本东京国立博物馆收藏）

图 4–24 唐三彩仕女俑（新疆吐鲁番阿斯塔那 187 号墓出土）

图 4–25 彩绘陶女俑（新疆吐鲁番阿斯塔那 187 号墓出土）

图 4–26 彩绘翻领胡装女性俑（陕西咸阳边防村出土）

图 4–27 乐舞屏风所绘女子像（新疆吐鲁番阿斯塔那高昌左卫张雄之孙张礼臣墓出土）

图 4–28 彩绘陶戴胡帽骑马女俑（陕西咸阳礼泉县李贞墓出土）

二、女效男装

在唐代以前，男女之间在服饰和服制上有着不可逾越的界限。“女效男装”为社会制度和礼仪规范所不容。跨越传统着装界限的唐代女性，却以身着男装、跃马扬鞭为一时风尚。西安的唐墓中就出土了许多身穿圆领袍，腰系銙带的彩绘女俑（图4–29、图4–30）。此外，西安昭陵唐墓壁画（图4–31）、《虢国夫人游春图》和《挥扇仕女图》等中皆有这种靓妆露面、无复障蔽的女子。“女效男装”现象在唐诗中也多有褒奖，如李贺《河南府试十二月乐词·三月》云：“军妆宫妓扫蛾浅”，司空图《剑器》云：“空教女子爱军装”。女子穿戎装，于秀美俏丽之中，别具一种英姿飒爽的气质。

图 4–29　彩绘陶男装女立俑（陕西西安郭家滩出土）

图 4–30　彩绘陶男装女立俑（陕西西安金乡县主墓出土）

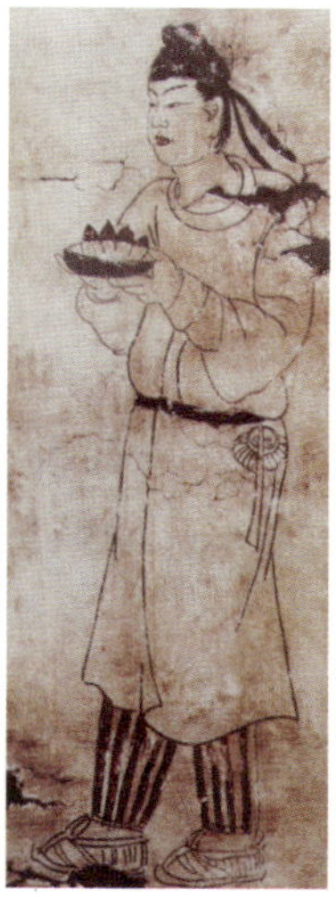

图 4–31　陕西西安昭陵唐墓壁画中身穿男装的侍女

图 4–32　唐三彩釉陶女立俑（陕西富平节愍太子墓出土，陕西省考古研究院藏）

图 4–33　唐代周昉绘《内人双陆图》局部

图 4–34　《捣练图》局部

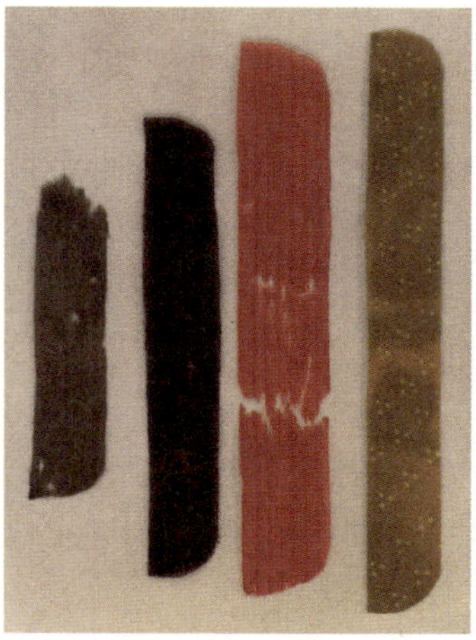

图 4–35　唐代彩色罗帔帛（新疆吐鲁番阿斯塔那张雄夫妇墓出土）

三、帔子

谈及唐代女性着装时尚，帔子是最不可缺少的元素。帔，也称帔子。唐代张鷟《游仙窟》诗云：“迎风帔子郁金香，照日裙裾石榴色。”帔一般用纱、罗等轻薄织物做成，而“罗帔掩丹虹”就是指用罗做成的帔。又由于帔的颜色多为红色，故古人也称帔为“红帔”，如唐三彩釉陶女立俑，该俑梳单刀翻髻，身穿小袖衫、高腰裙，肩披帔子（图 4–32）。帔子在唐代绘画作品中也多有表现（图 4–33、图 4–34）。新疆吐鲁番阿斯塔那张雄夫妇墓也出土有唐代彩色罗帔帛实物（图 4–35）。

时世女妆

中晚唐时期，社会风气以奢侈享乐为尚，女性妆饰追求浓艳丰盈。德宗时，长安妇女争妍斗艳已成生活大事，时世妆、椎髻、乌唇、八字眉等风靡一时。

唐代流行花钿、斜红、妆靥的面部装饰。花钿和妆靥合称花靥。花钿是用金箔片、珍珠、鱼鳃骨、鱼鳞、茶油花饼、黑光纸、螺钿壳及云母等制成。

发髻

唐代妇女发式主要分为髻、鬟、鬓三类。髻是一种实心的发式（图 4–36），有愁来髻、乐游髻、百合髻、归顺髻、盘恒髻、惊鹄髻、抛家髻、长乐髻、高髻、堕马髻、半翻髻等。鬟是一种环状而又中空的发髻，有云鬟、高鬟、短鬟、双鬟、垂鬟等。双鬟有分开梳在头顶两侧，也有梳在头顶正中，还有如神仙一般的“双环望仙髻”。

图 4–36 唐代女性发髻式样

图 4–37 唐代鎏金刻花摩羯纹莲叶纹银钗（陕西省博物馆藏）

簪头镂空五朵盛开的菊花，花朵间枝叶缠绕。

图 4–38 唐代鎏金菊花纹银钗（陕西省博物馆藏）

图 4–39 鎏金银花钗（浙江长兴唐墓出土）

高23厘米，在鎏金钗股上，以金丝镶嵌玉片，制成一对展开的蝴蝶翅膀。蝶翼之下和钗梁顶端也有以银丝编成的缀饰。

图 4–40 唐代金镶玉四蝶银步摇花钗（安徽合肥西郊南唐墓出土）

花钗

唐代流行一种与步摇相似的花钗饰品（图 4–37 ～图 4–40）。钗首制成鸟雀状，雀口衔珠，步行摇颤，倍增韵致。白居易《长恨歌》中载有“花钿委地无人收，翠翘金雀玉搔头。”佩戴花钗者非贵即富，因而制作工艺极尽奢华。通常而言，一副花钗纹样相同，簪戴时左右相望地插于发髻两侧。

半臂

半臂，又称半袖，是一种套在长袖衣外面，长度齐腰的短袖上衣（图 4–41）。唐高祖李渊将长袖衣剪成短袖半臂，引得世人，尤其是女性，竞相穿着，如韩琮《公子行》诗云：“紫袖长衫色，银蝉半臂花。”唐代女子半臂有与现代 T 恤相似的紧身袒胸、短袖套头的式样，也有用小带子在胸部系住的对襟翻领或无领的式样（见图 4–42）。中唐以后，半臂的穿用日趋少见。原因在于初唐女装流行窄身小袖、紧贴身体的式样，这种造型正适合穿着半臂，而盛唐流行博衣大袖的式样，不适合套穿窄小的半臂。

图 4–41　唐代红色织锦半臂和唐代加缬罗半臂（日本正仓院藏）

图 4–42　身穿半臂的唐代女性形象（陕西乾县唐永泰公主墓壁画）

图 4–43　彩绘刺绣靴（新疆尉犁营盘出土）

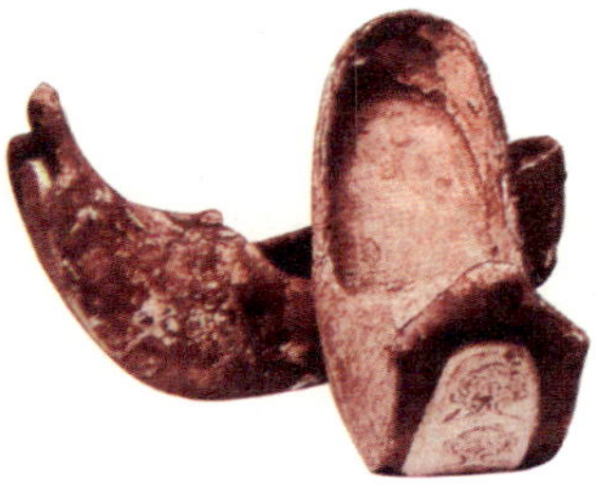

图 4–44　唐代棕色绣花高墙绢履（新疆吐鲁番阿斯塔那墓出土）

图 4–45　唐代变体宝相花纹云头锦履（新疆吐鲁番阿斯塔那 381 号墓出土）

高墙履

至隋代，袴褶服着靴已成定制。唐承隋制，无论贵贱高低均可穿靴。唐中期，上层社会女性流行穿软底镂空锦靴，与翻领、小袖袍服和条纹袴配套穿用。李白《对酒》云：“青黛画眉红锦靴”，其实物如新疆尉犁县营盘出土的彩绘刺绣靴（图 4–43），皮底麻面，内衬柔软保暖的毛织物。

唐代女子足服流行鞋尖上耸的高墙履（图 4–44）和重台履，如元稹《梦游春七十韵》诗云：“丛梳百叶髻，金蹙重台屦。”重台履也称丛头鞋，唐代和凝《采桑子 · 蝤蛴领上诃梨子》词云：“丛头鞋子红编细，裙窣金丝。”

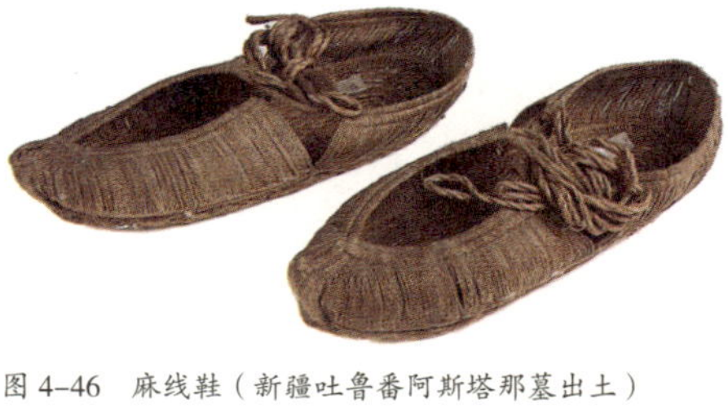

图 4–46　麻线鞋（新疆吐鲁番阿斯塔那墓出土）

唐代还流行鞋尖相对平缓的云头履，王涯《宫词三十首》云：“春来新插翠云钗，尚著云头踏殿鞋。”实物如新疆吐鲁番阿斯塔那 381 号墓出土的唐代变体宝相花纹云头锦履（图 4–45），浅色斜纹变体宝相花锦面，鞋尖内蓄棕草以同色锦扎翻卷云头。唐代还流行用麻线做的麻线鞋（图 4–46）。

第二节 西方：拜占庭

中世纪的两大风格——哥特式和拜占庭式，一东一西闪耀在欧亚大陆。

自从 395 年罗马帝国分裂后，西罗马帝国因北方日耳曼人的入侵于 476 年灭亡，但东罗马帝国的拜占庭文化却在整个中世纪一千余年的时间里，集古罗马文化、东方文化和基督教文化于一体，对同时期以及后世的西欧文明影响甚大。

拜占庭帝国分为前后两个时期，以 6 世纪查士丁尼（Justiniaus，527—565 年在位）统治时期为鼎盛期。前期（4—6 世纪），奴隶制逐渐瓦解，与此同时，封建主义作为新兴势力发展起来。查士丁尼王朝时期，独特的拜占庭文化被来朝拜进贡的西欧贵族带回西欧，影响着西欧文明的进程。

拜占庭文化是希腊、罗马的古典理念、东方的神秘主义和新兴基督教文化这三种完全异质的文化的混合物。拜占庭发达的染织业，是以华美著称的拜占庭文化的重要组成部分（图 4–47）。6 世纪中叶以前，中东一带通过丝绸之路与中国进行着频繁的交流，进口了大量中国丝绸。552 年，查士丁尼派遣两位熟悉东方情形的基督教徒远赴中国，将当时中国对外保密且禁运的蚕卵藏在竹杖中偷运回君士坦丁堡，从而使生丝的生产在拜占庭兴起，加速了拜占庭的经济繁荣。同时，查士丁尼又从叙利亚的纺织工匠那里引进机杼和织花技术，使拜占庭的丝织业迅速发达起来。

拜占庭初期服装延续着古罗马帝国末期式样。随着基督教文化的普及，拜占庭服装逐渐失去了古罗马服装的自然之美，造型变得平整而宽大，人体也被遮蔽在服装之下不再显露（图 4–48）。从中国进口的大量丝绸以及拜占庭染织业的不断发展，使人们有机会使用更为华丽和珍贵的丝织品。人们竭尽所能地增加纺织面料的外观效果，如金银线混织的中古织金锦（Samite），将宝石和珍珠织入织锦，这使服装的表现重点从古希腊、古罗马强调自然的人体意识转变为注重衣料的质地、色彩和装饰上，由此形成了西方服装史上的第一次“奢华时代”。拜占庭人相信，上天的力量显示在皇帝和教会的金银珠宝上，皇宫和教堂愈是豪华，就愈能证明每位基督徒所期盼的来世会成真。这种观念让拜占庭发展出了精湛的金银制品工艺技术。

图 4–47 拜占庭服饰

图 4–48 拜占庭时期人物服饰

达尔马提卡（Dalmatica）

拜占庭时期的主要服装是从罗马帝国末期与基督教一起出现和普及的达尔玛提卡（Dalmatica），它取代了托加（Toga）和帕拉（Palla）等款式。藏于罗马圣彼得大教堂的查理曼大帝的两件达尔玛提卡（Dalmatica）实物，工艺非常精美（图 4–49）。根据拜占庭的传统，大主教袍上的图案为圣像及宗教仪式场景。除了人物，大主教袍上还装饰有精美的花卉纹样（图 4–50）。

与贵族服装的奢华相反，普通劳动阶层的服装样式朴素简单，没有任何装饰。妇女们不施装扮，反将珠宝献给教会，自己平时穿着白色宽大长衫与连袖外套，素净淡雅，然而她们的丧服却又色彩鲜艳，用以祝愿逝者在来世能有幸福的生活。

391年，狄奥多西大帝把基督教定为国教后，具有宗教色彩的达尔玛提卡（Dalmatica）开始在罗马市民中普及。这是一种无性别和等级区分的日常服装。罗马市民把布料裁成十字形，中间挖洞作贯头的领口，袖下和体侧缝合。达尔玛提卡（Dalmatica）的肩部到下摆装饰着两条紫红色的纵向条饰克拉比（Clavi）（图4–51）。克拉比（Clavi）最早使用于古希腊服装，曾是贵族身份和地位的标志（图4–52）。到了拜占庭时期，克拉比（Clavi）作为基督血的象征，成为一种单纯的宗教装饰，不再具有先前的身份和等级象征，任何人都可以使用。

图 4–49　罗马圣彼得大教堂查理曼大帝的 Dalmatica

图 4–50　身穿花卉纹样 Dalmatica 的主教

图 4–51　拜占庭壁画中的 Dalmatica

图 4–52　古希腊陶器上身穿装饰着 Clavi 的 Dalmatica 的女性（美国纽约大都会艺术博物馆藏）

帕鲁达门托姆（Paludamentum）

拜占庭时期最具代表性的外衣是方形大斗篷帕鲁达门托姆（Paludamentum），其面料为毛织物，通常染成紫色、红色或白色。帕鲁达门托姆（Paludamentum）在罗马时代被披在左肩，右肩处用别针固定；在拜占庭时代变长为梯形，穿着者在胸前缝一块四边形的装饰布。其样式如圣维塔列教堂的镶嵌画《查士丁尼及其随从》（图 4–53）和《提奥多拉与随从》（图 4–54）中的人物穿的袍服。在《查士丁尼及其随从》中心位置的是身穿紫红色长袍的查士丁尼大帝，他的右边是两个表情谦恭的贵族和五个手拿武器的年轻侍卫，左边是大主教马克西米尔，再左边是两个助祭者，一个手里拿着精心装饰的《圣经》，一个提着教会中使用的油灯。

拜占庭时期大斗篷帕鲁达门托姆（Paludamentum）实物如罗杰二世的加冕丝质长袍，上面饰有珍珠与金线（图 4–55）。此时，帕鲁达门托姆（Paludamentum）的另一个特点是在斗篷的前面和后面各有一个方形或长方形的绣饰（图 4–56），称为塔布里昂（Tablion）。塔布里昂（Tablion）类似中国明清时期官员常服上的“补子”。

较之男装，拜占庭女装更具东方风格。6 世纪以后，罗马女性们穿的帕拉（Palla）逐渐变窄，称为帕留姆（Pallium），与达尔玛提卡（Dalmatica）一起作为外出服饰使用。到拜占庭时代，帕留姆（Pallium）演变为宽 15 ～ 20 厘米的表面有刺绣或宝石装饰的带状物，时称罗拉（Lorum，图 4–57）。

图 4–53 《查士丁尼及其随从》

图 4–54 《提奥多拉与随从》

图 4–55 罗杰二世的加冕长袍

图 4–56 拜占庭时期的大斗篷 Paludamentum

图 4–57 拜占庭时期的 Lorum

第五章　公元 11—13 世纪

第一节　中国：宋代

北宋：960—1127 年
南宋：1127—1279 年

960 年，赵匡胤统一全国，建立宋朝，从而结束了晚唐五代的混乱局面。宋王朝立国 319 年，其间以 1127 年“靖康之耻”为界，分为前、后两个时期。此前为北宋，定都汴京（现河南开封）；此后为南宋，定都临安（现浙江杭州）。

北宋时期，为了建立一个高度集权专制的封建王朝，宋王朝吸取了唐末藩镇割据、节度使拥兵自重的教训，实行“修文偃武”的政策，进而削弱军人权势，转而大力推行科举制度，士大夫官僚阶层崛起而取代了士族豪门，形成了文人治国的格局。宋代商品经济的繁盛和城市建设的发展，使宋人的物质生活和审美品位得到了极大的丰富和提升（图 5-1、图 5-2）。

宋代从事纺织作坊有官营和私营之分，广大农村的劳动妇女也都纺织布帛。北宋丝织以两浙、川蜀地区最为发达。开封设绫锦院，为皇室贵族织造高级织品。北宋和南宋棉花种植和桑蚕业遍及各地，织机遍及城乡。当时，丝织品已达 100 多种，纱、罗、绢、锦、绮、绫等品类都出现了许多名品，如单州薄缣望之若雾，亳州轻纱更是有名。宋代缝纫工具，如剪刀、熨斗和制针等工具也都有所发展。

以服为纲

宋初服制承袭五代旧规，后循唐制，上自皇帝、皇太子、诸王以及各级品官，下及于吏庶等。因议论繁多，宋政府曾多次更改和修订服饰制度。期间北宋初期太祖时期（960—976 年）及仁宗景祐二年（1035 年）、神宗元丰四年（1081 年）、徽宗大观四年（1110 年）和政和年间（1111—1117 年），及高宗绍兴四年（1134 年）的制度较详。宋代官服可分为祭服、朝服、公服和时服等。除祭祀朝会之外，百官公服为袍衫，并以颜色区别等级。其色有紫、绯、朱、绿、青不等。与前代群臣冠服，大多采用“以冠统服”的模式不同，自宋代开始变为“以服为纲”的模式。

图 5-1　宋代张择端绘《清明上河图》局部

图 5-2　宋代《文会图》局部

天子之服

宋代天子之服有七种，一曰大裘冕，二曰衮冕，三曰通天冠、绛纱袍，四曰履袍，五曰衫袍，六曰窄袍，七曰御阅服（戎服）。

通天冠服为皇帝大祭祀致斋、正旦冬至五月朔大朝会、大册命、亲耕籍田时穿用（图5-3）。首服通天冠，二十四梁。绛纱袍，绛纱裙，蔽膝如袍饰，白纱中单，方心曲领。白袜，黑舄，佩绶如衮。

履袍为皇帝郊祀、明堂，诣宫、宿庙、进膳、上寿两宫及端门肆赦时穿用。头戴折上巾，身穿绛罗袍，通犀金玉带。系履时称履袍，服靴时则称靴袍。履、靴皆用黑革，其式样如《宋仁宗坐像》（图5-4）。

衫袍为皇帝大宴时穿用，头戴折上巾，身穿淡黄袍衫，玉装红束带，皂文靴。自隋代始，天子的常服已为赤黄、浅黄袍衫、九还带、六合靴。宋代承袭，其式样如《宋太祖赵匡胤像》（图5-5）和《宋太宗立像》。

窄袍为皇帝便坐视事时穿用。头戴皂纱折上巾或乌纱帽。身穿红袍或褙子，系通犀金玉环带，圆领袍里面多穿交领袍。

方心曲领，是套在朝服交领之上的用白罗制成的项饰，为宋代朝服最为显著的特征之一。其意在附会天圆地方之说，如宋代《女孝经图》中的帝王（图5-6）和明代《三才图会》中的方心曲领（图5-7）。

图5-3 《宋宣祖通天冠服像》（南薰殿旧藏）

图5-4 《宋仁宗坐像》（南薰殿旧藏）

图5-5 《宋太祖赵匡胤像》（南薰殿旧藏）

图5-6 宋代《女孝经图》中的帝王像

图5-7 《三才图会》中的方心曲领

诸臣服制

宋代诸臣礼服有祭服、朝服和公服三类。

祭服有衮冕九旒，毳冕七旒，玄冕五旒，平冕无旒。中兴之后，宋代诸臣祭服定为四等：一曰鷩冕，八旒；二曰毳冕，六旒；三曰絺冕，四旒；四曰玄冕，无旒。

宋代诸臣朝服有三式：进贤冠（图5-8）、貂蝉冠和獬豸冠，皆朱衣朱裳。宋代进贤冠式样与唐代变化较大，其冠体的前屋造型趋于饱满，且附梁于其上，后衬一横向山墙，从后面向前包，与汉代纳言式样成反例。其额前，又衬额花装饰（图5-9）。其式样一直沿袭至明代（图5-10）。

自宋以降，进贤冠外面还罩有笼巾，改称"貂蝉冠"（图5-11、图5-12），织藤漆之，形正方，如平巾帻。饰以银，前有银花，上缀玳瑁蝉，左右各为三小蝉，玉鼻，左插貂尾。三公、亲王侍祠大朝会，则加于进贤冠而服之。官品照等级于笼巾下附有不同卷梁装饰，耳侧或附一雉尾或貂尾。

在宋代，朝服被称为具服，公服从省，以圆领大袖常服代替。宋因唐制，三品以上服紫，五品以上服朱，七品以上服绿，九品以上服青。其制，曲领大袖，下施横襕，束以革带，幞头，乌皮靴。自王公至一命之士，通服之。

图5-8　张宗子绘《越中三不朽图赞》中头戴七梁进贤冠的翰林院侍讲

图5-9　明代五梁进贤冠（山东孔府旧藏）

图5-10　《瑞应图》中的宋代官员

图5-11　宋代范仲淹像中的貂蝉冠

图5-12　明代官员头戴貂蝉冠

图5-13　圆领素罗大袖袍（浙江黄岩南宋赵伯沄墓出土）

素罗大袖袍

宋代官员朝服如素罗大袖袍（图5-13），墓主赵伯澐乃宋太祖赵匡胤七世孙，衣长115厘米，通袖长239厘米，袖宽95厘米，袖根宽46厘米，胸宽72厘米，下摆宽96厘米，圆领右衽，袖身宽大，采用对幅式断腰裁剪，衣身分成上衣下裳再连属。肩部与腋下有扣和扣襻。其面料为四经绞罗，单层无衬里。

道袍

宋代服饰，以存古风为尚，道袍、襕衫、直裰、鹤氅等款式流行。

道袍本为释道之服，宋、明时期广泛穿着，流行度极高，上至皇帝、官员，下至士庶男子皆可穿着（图 5–14），宋代道袍多以素布为主，交领右衽、两袖宽博，领、袖、襟、摆处有深色缘边，下长过膝，两侧没有开衩，形象如宋人刘松年《听琴图》中的人物服装（图 5–15）。

直裰

直裰，也写作“直掇”。最初多用作僧道之服，后为文人士人所穿的男服。其式样为对襟大袖（宋代为直袖，明代为收祛的琵琶袖），衣缘四周镶有黑边，其后背有一条直通到底的中缝，前襟上也有一条中缝。其最初多用作僧人和道士之服（图 5–16）。到了元明时期，直裰形制有所变异，通常用纱縠、绫罗、绸缎及苎麻织物制成，大襟交领，下长过膝，多用于士庶男子。

图 5–14 明代蓝色暗花纱道袍（山东曲阜孔府旧藏）

图 5–15 宋代赵佶绘《听琴图》中身穿道袍的人物形象

图 5–16 宋代赵佶绘《听琴图》中身穿直裰的弹琴者形象

图 5–17 宋代《青城像》

图 5–18 宋代赵佶绘《听琴图》中身穿襕袍的男子形象

图 5–19 明代《三才图会》襕衫

图 5–20 穿蓝衫的明代士人画像

襕衫

襕衫乃古代士人之服，圆领或交领、宋代无缘饰，明代有缘饰，下摆有一横襕（缝缀或织成的纹样），以示上衣下裳之旧制。其制始于北周，兴于唐代，流行于宋明。据《宋史・舆服志》载：“襕衫以白细布为之，圆领大袖，下施横襕为裳，腰间有襞积，进士、国子生、州县生服之。”宋代襕衫（图 5–17、图 5–18）接近于官员服制，为仕者燕居、告老还乡或低级吏人服用。

明代襕衫（图 5–19）还出现了以蓝色布料制作、没有膝襕的“蓝衫”。其下摆处有深色宽缘，以衣缘代替膝襕的象征意义（图 5–20）。

图 5-21　《历代帝后像》中的宋代皇后（南薰殿旧藏）

图 5-23　罗绣花霞帔（福建福州南宋黄昇墓出土）以素罗制成，裁制成条状，左右各一飘带，通体绣以花纹。

图 5-24　扁圆浮雕双凤金帔坠（福建福州南宋黄昇墓出土）

图 5-22　褐黄色罗镶花边大袖衣（福建福州南宋黄昇墓出土）

图 5-25　《宋宣祖皇后坐像》（南薰殿藏）

皇后礼服

宋代皇后、妃有祎衣、褕翟、鞠衣、朱衣和礼衣。祎衣（皇后受册、朝谒景灵宫服之，图 5-21），九龙四凤冠，深青色祎衣，青纱中单，蔽膝随裳色，大带随衣色。褕翟（皇后及皇太子妃受册服之），九翚四凤冠，青罗衣，素纱中单，蔽膝随裳色，大带随衣色。鞠衣（皇后亲蚕服之），黄罗衣，蔽膝、大带、革舄随衣色，余同祎衣，唯无翟纹。朱衣（皇后朝谒圣容及乘辇服之），绯罗衣，蔽膝、革带、佩绶，履随裳色。礼衣（皇后宴见宾客服之），钿钗礼衣，服通用杂色。此外还有常服（皇后常服为命妇的礼服），真红色大袖衣（图 5-22），红罗长裙，红霞帔，红罗褙子。

命妇礼服

宋代命妇随男子官服而区分等级，受册、从蚕典礼礼服：青罗绣翟鸟纹翟衣，素纱中单，蔽膝同裳色。花钗冠，两博鬓，一品花钗九株，宝钿数同花数，衣绣翟鸟九只；二品花钗八株，衣绣翟鸟八只；三品花钗七株，衣绣翟鸟七只；四品花钗六株，衣绣翟鸟六只；五品花钗五株，衣绣翟鸟五只。

帔帛至宋代演变成带状披肩而双垂的“霞帔”（图 5-23）。霞帔下端还系着雕饰凤鸟纹样的鸡心形帔坠（图 5-24）。其佩戴形式应如《宋宣祖皇后坐像》中的式样（图 5-25）。

褙子

褙子，也写作“背子”，是宋代最具代表性，且上至皇后、贵妃、命妇，下至平民、侍从、奴婢，以及优伶、乐人不分等级与尊卑的通用性服装款式。

宋代褙子式样的主要特征为瘦身窄袖、对襟直领、加缝生色领、腋下开衩和不施衿纽（图 5–26、图 5–27）。与唐朝女子骑马射猎、英姿飒爽的风格不同，宋朝女子追求的是一种朴素雅致、含而不露而又风情万种的小家碧玉之美。宋人绘《荷亭戏婴图》《瑶台步月图》（图 5–28）可见修长清秀的宋代美女。

宋代褙子的前襟不施纽襻，两裾离异，谓之“不制衿”，且腋下侧缝缀有带子，垂而不结仅作装饰，意义是模仿古代中单交带的形式，有“好古存旧”之意。

图 5–26　紫灰色绉纱镶花边窄袖褙子（福建福州南宋黄昇墓出土）

图 5–27　印金罗襟折枝花纹罗衫实物（江西德安南宋周氏墓出土）

图 5–28　宋代刘宗古绘《瑶台步月图》

图 5–29　浅褐色绉纱镶花边窄袖衫（福建福州南宋黄昇墓出土）

图 5–30　深烟色牡丹花罗背心（福建福州南宋黄昇墓出土）

图 5–31　深烟色罗牡丹芙蓉花缠枝纹样

紫衫

除了褙子、大袖衫，宋代女服还流行衣身较短的“紫衫”，即紫颜色的短衫，前后开衩，便于乘骑。隋时皇帝侍从服用，唐代多用于武士，宋时为军校之服。据宋代袁文《甕牖闲评》记载，南宋以后，被定为军校之服；皂隶走卒也多着之。及至南宋，战事频仍，文官士人竞着紫衫，以便从戎（图 5–29）。

背心

宋代是背心的流行时期，男女不论尊卑均喜穿着。其式样为直线裁成，呈长方形，对襟直领，衣身长至腰际，下摆两侧各开一衩，两襟之间不用搭襻，也无纽扣，穿时任其敞开，如福建福州南宋黄昇墓出土的深烟色花罗无袖衫（图 5–30），后人称之为背心。这些纱罗面料多有缠枝暗花纹样（图 5–31）。

女裙

宋代妇女下身多穿裙和裤（图 5-32）。宋代女裙多存唐代遗制，时兴“石榴”、“千褶”和“百迭”等名目。言“千褶”者，为细裥女裙。千褶裙衣长托地，以五色轻纱为之，周身折裥，裥多而密。

宋代还有一种前后开衩的裙子，乘骑方便，时称“旋裙”。宋初流行于京都青楼女子，后影响至士庶间。宋理宗时，宫廷嫔妃时兴穿一种前后相掩（不缝合）的裙式，因走起路来裙裾扫地，被称为“赶上裙”（图5-33）。因其前后都可开合，故被保守的人士视为“服妖”。与上身穿的衫一样，宋代女裙也流行使用轻薄质料，时有“薄罗衫子薄罗裙”之称。

宋代女裙中还有一种叫腰裙的短裙。其长者下不掩踝，短者不及膝盖。从文献记载看，短裙虽也有用于礼仪场合的时候，但一般多用于日常生活。短裙一般多用作衬里，俗谓“衬裙”。

图 5-32　褐色罗印花千褶裙（福建福州南宋黄昇墓出土）

图 5-33　褐色牡丹花罗镶花边裙（福建福州南宋黄昇墓出土）

图 5-34　黄地小杂花金锦夹吊敦实物（黑龙江哈尔滨阿城金墓出土）

图 5-35　宋人绘《杂剧图》

图 5-36　翘头绣花女鞋（福建福州南宋黄昇墓出土）

图 5-37　翘头绣花女鞋（浙江兰溪南宋墓出土）

吊敦膝裤

宋代裤子质料多以纱、罗、绢、绸、绮、绫为主，颜色以驼、黄、棕、褐等色为主。

宋代曾流行契丹服饰吊敦（袜裤），即一种只有两条裤管的内裤。吊敦上沿呈内低外高形态，与人体腿部形态吻合。其裤腿上部有长带，将其系于膝盖下部（图 5-34）。其穿着式样如宋人所绘的《杂剧图》形象（图 5-35）。

乌靴金莲

五代以前男女的鞋子是同一形制。从宋代开始，中国女性流行用布帛把脚缠裹起来。宋代缠足与后世的三寸金莲有所区别。据史籍记载，宋代的缠足是把脚裹得“纤直”但不弓弯，时称“快上马”。鞋底前尖、两色合成的鞋称“错到底”（图 5-36、图 5-37）。宋代风俗画家王居正《纺车图》和宋人绘《杂剧图》中也有着缠足形象。

第二节 西方：罗马式时代

罗马式艺术即七八世纪以后，在西欧兴起的艺术。罗马式艺术在 11 世纪成熟，12 世纪达到巅峰，可以说是当时“泛欧洲”的艺术潮流，其影响力延续至 13 世纪。欧洲历史上一般把 11 至 12 世纪称为“罗马式时代”（Romanesque）。这是欧洲中世纪出现的两大国际性时代中的第一个时代。罗马式艺术于 1075 至 1125 年间达到高潮，12 世纪中叶以后被哥特式艺术取代。

图 5-38 罗马式时代女式 Bliaut 和 Chains

男女同服

罗马式时代的服装是日耳曼人吸收基督教和罗马文化后，逐渐形成独自的服装文化的过程。它一方面在形式上继承了古罗马和拜占庭的宽衣、斗篷、风帽和面纱，另一方面则保留了日耳曼系腰带的丘尼卡（Tunic）窄衣样式。

罗马式时代的服装特征是男女同型，除男子穿裤外，几乎没有明显的服饰性别差异。其基本品种有内衣鲜兹（Chains）、外衣布里奥（Bliaut）、斗篷曼特尔（Mantle）。

内衣鲜兹（Chains）和外袍布里奥（Bliaut）都是长长的筒形丘尼卡（Tunic）式衣服。鲜兹（Chains）是白色麻织物的内衣，袖子为窄长紧身的造型，袖口装饰着精美的刺绣和带子，领口多以数排丈绳或金银线滚边作缘饰，衣长及地（图 5-38、图 5-39）。

图 5-39 罗马式时代男式 Bliaut 和 Chains

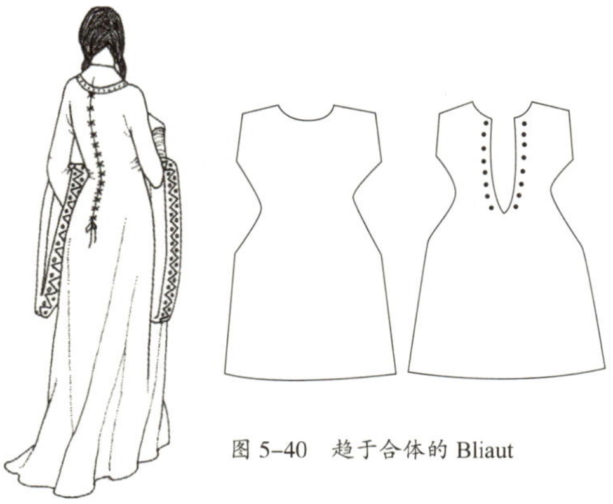

图 5-40 趋于合体的 Bliaut

图 5-41 有许多纵向褶皱的 Bliaut

布里奥（Bliaut）

布里奥（Bliaut）是12至13世纪欧洲男女穿着的束腰长袍，一般选用丝绸织物或毛织物制作。它下摆、肩、胸、背宽松，腰节处收紧，袖口进行斜裁，七分袖或八分袖，袖口呈喇叭状。为了使造型合体，人们将前片和后片按人体的腰部曲线裁剪，并在后中线竖向破缝，在破缝两侧挖汽眼，并用绳带系紧。这样使布里奥（Bliaut）出现了许多横向的衣褶（图5-40）。其袖根到肘部紧身，肘部以下的袖口突然变大，长者甚至垂地。为了方便，人们还将袖口打结。裙子下摆呈扇形，长长地拖在地上，盖住穿者的双脚。

女装打破了直线剪裁方式，首创性地在侧摆处加入三角布来增大臀围量以及下摆量，这样就迈出了服装合体性的第一步，并由此形成了服装表面丰富的纵向衣褶（图5-41）。

徽章

11至12世纪，西欧的金属冶炼和制造工艺有了很大的提高。西欧骑士的头盔和铠甲由原来的半裸型向封闭型转变，骑士的脸部与躯体几乎完全被遮住。为了区分穿者的身份，就在盾、旗帜、胴衣、外套、马披、马鞍、帐篷及其他器具上都画上、刺绣或雕刻一些易于识别的纹样(图5-42)。13世纪末，法国女装也应用了这种纹样（图5-43）。到了14世纪，甚至连一般市民和农民也流行使用家徽来装饰服装。当时的人们崇尚骑士精神，热衷于建功立业，彰显家世，徽章也被看成家族标记，显示出主人高贵的社会地位，并进而与采邑、封地乃至家族联系起来，最终成为一种荣耀和不可侵犯的世袭符号。

图 5-42 中世纪武士盾牌和服装上的徽章图案

图 5-43 用徽章装饰的中世纪女装

第六章　公元 13—15 世纪

第一节　中国：元代

元代：1206—1368 年

1206 年，元太祖铁木真（尊称“成吉思汗”，图 6–1）建国；1232 年，窝阔台汗（图 6–2）南下灭金；1260 年，铁木真的孙子元世祖忽必烈登上汗位（图 6–3）；1271 年，迁都大都（现北京），改国号为元。随后，在经过了 8 年的努力后，于 1279 年灭南宋，结束了从五代到南宋 370 多年多政权并立的局面，建立了统一的大帝国元朝。

建国以后，元朝加强了欧亚大陆之间的贸易和文化交流活动，除汉文化外，还受到吐蕃喇嘛教文化、中亚伊斯兰文化，乃至欧洲基督教文化的影响。因此，元朝社会具有地域辽阔、种族混杂，农耕文化与草原文化、佛教文化与伊斯兰文化以及欧洲基督教文化相互融合的特点。这就造成了元朝服饰的多样化。

元朝统治者曾经把中国的各种技工集中起来，再安置在中国各省以及中国以外势力所及的地方。由于生长于草原环境，蒙古人所爱用的色彩，不论是建筑、衣服、绘画等，都偏重于浓重鲜艳，以打破自然色调上的沉寂。蒙元王朝，以白色与蓝色为贵，以红色为尚。帝王的旌旗、仪仗、帷幕、衣物以白色居多。蒙古族尚红与萨满教信仰“万事万物都是被火所净化”有关，尊崇火从而导致对红色的推崇。在当时，禁限的九种色彩中，便有红白闪色、鸡冠紫、栀红、胭脂红、真紫五种与红色有关。百官公服，只准五品以上服紫、七品以上服绯。汉族和尚不得衣红，只有吐蕃僧人才可，甚至规定除寺观、孔庙外，大门不准髹红色。

元代民间禁止穿赭黄、柳芳绿、红白闪色、迎霜色（褐色）、鸡头紫、栀子红、胭脂红等颜色。因此，民间服饰只好向灰褐色系发展。

图 6–1　元太祖铁木真像

图 6–2　元太宗窝阔台汗像

图 6–3　元世祖忽必烈像

图 6–4　牵马俑（灰陶，俑高 34 厘米，马高 37 厘米，陕西西安户县元贺氏墓出土，陕西博物馆藏）

图 6–5　元代绘画中的诈玛宴

一色质孙

元代版图非常大，横跨欧亚两大洲。蒙古在进入中原以前从事比较单纯的游牧和狩猎经济（图 6–4），对汉族农业文明接触较少。

蒙元时期蒙古贵族流行穿质孙服。质孙，蒙语音译，亦作“只孙”“只逊”“济逊”“直孙”，其含义为“一色服”“一色衣”，为元代宫廷举行盛宴时，皇帝和百官共同穿着的礼服（图 6–5）。元代统治者每年要举行 13 次大朝会。每逢此时，帝王、大臣、亲信穿同一色的质孙服在大殿前用金杯按爵位、亲疏、辈分频频祝酒，场面壮观。

据《元史·舆服志》载，元天子质孙服：

冬之服凡十有一等，服纳石失、金锦也。怯绵里，翦茸也。则冠金锦暖帽。服大红（图 6–6）、桃红、紫蓝、绿宝里，则冠七宝重顶冠。服红黄粉皮，则冠红金答子暖帽。服白粉皮，则冠白金答子暖帽。服银鼠，则冠银鼠暖帽，其上并加银鼠比肩。

夏之服凡十有五等，服答纳都纳石矢，缀大珠于金锦。则冠宝顶金凤钹笠。服速不都纳石矢，缀小珠于金锦。则冠珠子卷云冠。服纳石矢，则帽亦如之。服大红珠宝里红毛子答纳，则冠绿边钹笠。服白毛子金丝宝里，则冠白藤宝贝帽。服驼褐毛子，则帽亦如之。服大红、绿、蓝、银褐、枣褐、金绣龙五色罗，则冠金凤顶笠，各随其服之色。服金龙青罗，则冠金凤顶漆纱冠。服珠子褐七宝珠龙答子，则冠黄牙忽宝贝珠子带后檐帽。服青速夫金丝阑子，则冠七宝漆纱带后檐帽。

图 6–6　元代刘贯道绘《元世祖出猎图》及局部

辫线袄

辫线袍（袄），俗称“辫线袄子”，也称“腰线袄子”。辫线袍产生于金代，大规模使用则在元代。《元史·舆服志》载：“辫线袄，制如窄袖衫，腰作辫线细褶。”其式样为圆领、紧袖、下摆宽大有许多褶裥。在《事林广记》中，步射总法插图射箭人物身穿腰部缝以宽阔围腰的袍服（图6–7）。其实物如元代滴珠奔鹿纹纳石矢辫线袍（图6–8）。有的辫线袍还在腰部中央钉有纽扣。辫线袍一直沿袭到明代，不仅没有随着大规模的服制变易而被淘汰，反而成了上层官吏的装束，连皇帝、大臣都穿着（图6–9）。

海青衣

海青衣，是指大袖衣，因衣袖宽博，如海青（亦称“海东青”，一种大海鸟）展翅，故名。语出唐代李白诗“翩翩舞广袖，似鸟海东来”，凡为大袖之衣，均可称之。元代宫廷画家刘贯道《元世祖出猎图》中就有身穿海青衣的人物形象（图6–10），实物如元代菱地飞鸟纹绫海青衣（图6–11）。其左右肩用钉金高绣法刺绣葵花形和一个底边宽24厘米的缠枝莲花纹样。

至清代，由于衣衫尚窄袖，海青大袖衣则成为僧服。徐珂《清稗类钞·服饰》载：“海青，今称僧尼之外衣也。然古时实以称普通衣服之广袖者。”

图6–7 《事林广记》中步射总法插图射箭人物

图6–10 元代刘贯道绘《元世祖出猎图》（局部）中身穿海青衣的人物形象

图6–8 元代滴珠奔鹿纹纳石矢辫线袄

图6–9 元代辫线袄实物

右衽交领，宽摆，袖略宽，肩部有一开口，手臂可从该处伸出。两袖离袖口16厘米处各有一个襻。两只长袖可在衣服背后离领子14厘米处的纽扣处反扣。袍服右襟处有33厘米开衩，便于骑乘。

图6–11 元代菱地飞鸟纹绫海青衣

褡胡

元代官员常在袍服外套一种半袖衣（图6–12），汉语称“比肩”，蒙语称“褡胡”“搭护”“绰子”“答忽”“褡护”。其形象如美国纽约大都会艺术博物馆藏元代织御容中元代皇帝像和内蒙古赤峰元代墓葬壁画中端坐女子穿的外衣（图6–13）。元人武汉臣《生金阁》载：“孩儿吃下这杯酒去，由于你添了一件绵搭护么？”显然，褡胡不仅有单夹，亦有内絮丝绵的。《元史·舆服志》中还有用银鼠皮制作的“比肩”。

蒙元被明朝取代后，虽然“诏复衣冠如唐制”，但仍有不少胡俗被沿袭，而褡胡便是其中之一。

图6–12　元代褡胡实物

图6–13　元代墓葬壁画中身穿褡胡的女子坐像

图6–14　湖塘明墓缠枝莲花缎比甲（常州博物馆藏）

图6–15　元代花绸背心（江苏无锡市郊元墓出土）

图6–16　明代月白暗花纱比甲（山东曲阜孔府藏）

图6–17　山西平遥双林寺彩塑仕女塑像

图6–18　《燕寝怡情》中身穿比甲的女性形象

比甲

除了半袖的比肩，元人还穿一种无袖的比甲。据《元史》载：“又制一衣，前有裳无衽，后长倍于前，亦去领袖，缀以两襻，名曰‘比甲’，以便弓马，时皆仿之。”明代比甲演变成无袖对襟、两侧开衩，衣长至膝的式样，如常州博物馆藏湖塘明墓缠枝莲花缎比甲（图6–14）和江苏无锡市郊元墓出土的花绸背心（图6–15）。

明代比甲大多为年轻妇女所穿，而且多流行在士庶妻女及奴婢之间（图6–16、图6–17）。到了清代，这种服装更加流行。明末清初《燕寝怡情》中有身穿比甲的女性形象（图6–18）。

到了清代，比甲更加流行。同时，比甲缩短衣身，称为坎肩或马甲。后来的马甲就是在此基础上经过加工改制而成的。

女袍

元代命妇衣服，一品至三品服浑金，四品、五品服金答子，六品以下惟服销金，并金纱答子。首饰，一品至三品许用金珠宝玉，四品、五品用金玉真珠，六品以下用金，惟耳环用珠玉。袍是元代女服中的主类。袍式多宽大，长可及地，宽衣长袖，袖口处窄小。如敦煌莫高窟中第 332 窟中元代身穿袍服的女供养人服装（图 6–19）、美国纽约大都会艺术博物馆藏元代织御容皇后像，其实物如元代团窠纹织金锦大袖袍（图 6–20）。

图 6–19　敦煌莫高窟第 332 窟中元代身穿袍服的女供养人

图 6–20　元代团窠纹织金锦大袖袍

图 6–21　加彩女俑（陕西西安曲江池元墓出土）

图 6–22　元代印金罗长袖衫（中国丝绸博物馆藏）

图 6–23　元代印金罗短袖衫（中国丝绸博物馆藏）

图 6–24　棕褐色四经绞素罗花鸟绣夹衫（内蒙古乌兰察布集宁路元代遗址出土）

短袖

元代女性习惯在长袖衫外套短袖（图 6–21）。其实物如中国丝绸博物馆藏元代印金罗长袖衫（图 6–22）和短袖衫（图 6–23）。长袖衫的袖子部位用金箔印有水波地麒麟纹样，领子和门襟处有方搭纹；短袖衫主体是小方搭纹，在其肩部和后背部分别有三角和方形装饰区，其内印有凤穿牡丹纹样。元代短袖实物如内蒙古乌兰察布集宁路元代遗址出土的棕褐色四经绞素罗花鸟绣夹衫（图 6–24）。这件夹衫为对襟直领，前襟长 60 厘米，通长 65.5 厘米，袖长 43 厘米，表面采用平绣针法，结合打籽针、辫针、戗针、鱼鳞针等针法，刺绣 99 个 5 ～ 8 厘米的图案，花型均为散点排列，时称“散搭子”。其题材有凤凰、野兔、双鱼、飞雁以及各种花卉纹样等。其中，最大的花形分布在两肩及前胸部分，一鹤伫立于水中央，一鹤飞翔于祥云之间，相望而呼应。鹤旁衬以水波、荷叶以及野菊、水草、云朵等，显现出一片生机勃勃的景象。衣服上有撑伞荡舟等人物故事图案，非常引人注目。

冬帽夏笠

元代男子公服多用幞头，平民百姓多用巾裹，又有“冬帽夏笠”的说法。

蒙元时期男子尚戴圆顶的“笠子帽”和方顶的“瓦楞帽”。笠子帽（图 6–25、图 6–26）是用四个大小相同的梯形毡片缝成帽身，再加缝帽顶。其式样有三种：宽檐和半球形圆顶；窄檐；四边帽檐向上折起的毡帽。瓦楞帽是用藤篾或牛马尾编结制成（图 6–27）。笠子帽和瓦楞帽都是古代北方游牧民族的传统帽饰。

元代皇室的帽子多镶宝石（图 6–28）。其品种中红宝石有四种：剌、避者达、昔剌泥、古木兰。绿宝石有三种：助把避、助木剌、撒卜泥。另外还有猫眼石、绿松石（称作甸子）、鸦鹘石等宝石。

元代和历代王朝一样对各色人等的服饰有详细规定。普通人戴前圆后方的笠帽。青楼掌柜头上戴的头巾必须为绿色，这就是“绿帽子”的来源。

罟罟冠

元代女子首服以“罟罟冠”最具特色（图 6–29、图 6–30）。罟罟冠上部以羽毛或树枝等物为之；中部呈上宽下窄的圆筒状；下部为冠筒与头部相连接的部分，早期为兜帽，晚期为抹额等。罟罟冠冠筒内胎最初用桦树皮，进入中原后以竹为骨架。冠筒外面裱覆的材料及装饰根据戴者的贫富及地位有所区别，贫者裱以褐，富者裱以绡、罗、绢、金帛等，并附珍珠宝石装饰。其实物如韩国藏元代方顶罟罟冠（图 6–31）。

蒙古族建立元朝进入中原地区后，冠顶的翎管与翎羽便很少使用了，如元帝后像、敦煌莫高窟和安西榆林窟元代壁画里的女供养人所戴的罟罟冠顶已经不见了翎管，较长的翎羽也被冠顶后面较短的“朵朵翎”所取代。这是因为在草原上生活时，高高的翎管和翎羽是贵族女性身份识别的标志，进入中原之后，其功能已被华丽的装饰所取代，因而只留下了象征性的朵朵翎。

图 6–25　笠子帽（甘肃漳县元代汪世显家族墓出土）

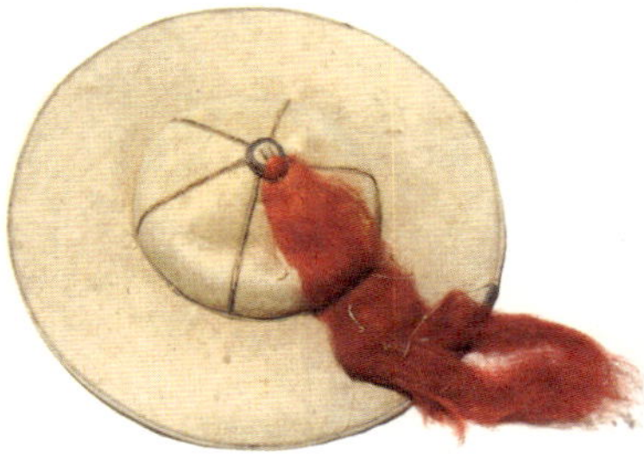

图 6–26　笠子帽（李雨来先生收藏）

图 6–27　元代瓦楞帽实物

图 6–28　头戴七重宝顶冠的元仁宗像

图 6–29　《元顺宗后塔济像》（台北故宫博物院藏）

图 6–30　元代织御容中的皇后像（美国纽约大都会艺术博物馆藏）

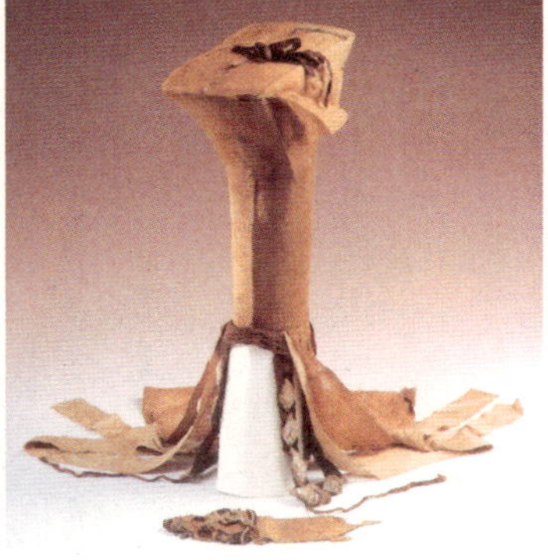

图 6–31　元代方顶罟罟冠实物（韩国藏）

第二节 西方：哥特式时代

哥特式艺术始于 12 世纪的法国，盛行于 13 世纪，至 14 世纪末期，其风格逐渐大众化，成为国际哥特风格，直至 15 世纪，因欧洲文艺复兴时代来临而迅速没落。不过，在北欧地区，这种风格仍延续了相当长的时间。

哥特式风格以高耸、阴森、诡异、恐怖为符号，呈现出夸张、不对称、奇特、轻盈、复杂和多装饰的风格式样，被广泛地运用在建筑、雕塑、绘画、文学、音乐、服装、字体等各个艺术领域。“哥特”是英语 Goth 的音译，原指代西欧日耳曼部族的哥特人。也有人说，Gothic 源于德语 Gotik，词源是 Gott，音译“哥特”（意为“上帝”），因此哥特式也可以理解为“接近上帝的”的意思。

科特（Cotte）

在哥特式初期，男女服装仍处在无性别区分的宽大筒形阶段。随后，罗马式布里奥（Bliaut）被新式外衣科特（Cotte）取代（图 6-32）。科特（Cotte）仍是一种男女同型的服饰。在正式场合或者外出时，人们在科特（Cotte）外罩一件修尔科（Surcot，图 6-33）。修尔科（Surcot）是一种装饰性的华丽套头筒形外衣。男女样式接近，有无袖、半袖、长袖式样。男式衣身较短，两侧开衩到胯部，便于行动。女式多为喇叭口的半长袖，可以露出里面的紧口袖，富有层次美，裙摆也很宽大，呈喇叭状，通常装饰有毛边。为了行走方便，人们常把前摆提起来，别进腰带，前面的堆褶使女性腹部像孕妇般凸起。当时还流行一种外衣叫希克拉斯（Cyclas）（图 6-34）。盛装时，人们还会披上超大的斗篷（图 6-35）。

图 6-32 中世纪 Cotte 实物

图 6-33 中世纪绘画中身穿 Surcot 的人物形象

图 6-34 中世纪 Cyclas

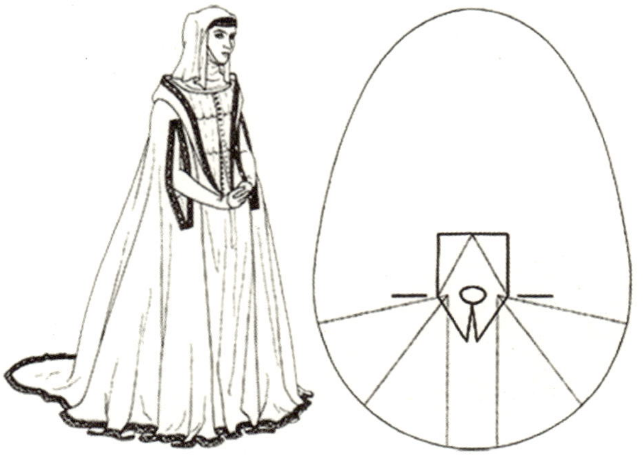

图 6-35 哥特式时代身穿斗篷的女子形象和斗篷裁剪图

中世纪女袍

中世纪时期，由于基督教神学的统治和禁欲主义的盛行，直线裁制的宽大衣袍抹杀了男女性别特征（图 6–36）。但人们对美的追求始终没有改变，到了中世纪末期的 13 至 14 世纪，女装从两个方面开始突破传统禁锢。首先，开大领口，更多地裸露女性性感部位——先是出现了横向扩大（领口为船形），增加颈部裸露，后来又出现纵向深开的 V 型领口，增加胸部裸露（图 6–37）。其次，是收紧腰身，以显露女性的腰臀之美。

图 6–36　中世纪初期禁欲主义的服装式样

图 6–37　中世纪穿深开 V 领服装的女子形象

科塔尔迪（Cote Hardi）

随着财富的增加，世俗兴趣的增强，欧洲人的自豪感和人生自由程度都有所提高。人们对人体的表现有了更高的要求。这一切在14世纪发生了改变。人们开始裸露肉体，将领口开大，袒露肩部和胸部。这与三维剪裁技术的应用相呼应，人性价值开始彰显，宗教色彩逐渐从服装上退位。这种倾向突出地表现在14世纪出现的外衣科塔尔迪（Cote Hardi，图6–40）上和无袖长袍萨科特（Surcote）上。

科塔尔迪（Cote Hardi）是一种连衣裙式外衣，源于意大利，14 世纪流行于西欧。科塔尔迪（Cote Hardi）运用了省的裁剪方法，使胸腰臀都非常合体贴身，突出女性的曲线美，在前中央处或者腋下置扣，或是系带子。它的领口很大，使得双肩袒露，是欧洲服饰史上第一次展现女性美的服装。

格陵兰长袍

人们借鉴按人体结构构成的欧洲铠甲（图 6–38）的立体造型技术，创制了格陵兰长袍（图 6–39）。它是一种上半身合身、下半身宽大的长袍。左右两侧及前后两面，从袖根到下摆共有 13 片衣片。为了使裙摆夸张，裙身插入许多条三角形布，诞生了真正意义上的三维立体服装。由于收省，胸部自然隆起而显出女性的曲线美。这种裁剪方法使罗马式时期服装的收腰意识在 13 世纪得到了发展，使服装结构由过去的二维空间向立体裁剪的三维空间方向发展。

图 6–38　欧洲中世纪时期盔甲

图 6–39　哥特时期格陵兰长袍款式和剪裁图

图 6–40　14 世纪身穿 Cote Hardi 的女子

萨科特（Surcote）

萨科特（Surcote）袖窿开得很深，而且前片比后片向里挖得更多，穿着时可以露出里面的衣服（图6–41）。它的前胸用宝石做成一排扣子，与穿在里面的科特(Cotte)袖子上的扣子相互呼应(图6–42)。

波尔普万（Pourpoint）

进入14世纪以后，为了增加服装的功用性，西方男性开始穿着称之为波尔普万（Pourpoint）的上衣和腿衣相结合的装束。波尔普万（Pourpoint）原来是穿在盔甲里面的上身衣服，后来转用为一般上衣（图6–43）。

图6–41 中世纪绘画中身穿Surcote的女子形象

图6–42 身穿Surcote的女子及其款式图

图6–43 Pourpoint实物

图6–44 中世纪人物画中的Chausses

肖斯（Chausses）

中世纪初，男女皆穿以丝绸、薄毛织物、细棉布等为材料做成的无裆袜肖斯（Chausses），英语称Hose。由于男子上衣已缩短至腰上部，肖斯（Chausses）需要用绳子与波尔普万（Pourpoint）的下摆系接固定（图6–44）。这段时期肖斯（Chausses）的左右裤腿是分开的，且流行左右不同色的穿法。有的肖斯（Chausses）保持袜子形状，在脚底还有单独缝合的皮革底，有的已进化为裤子状，长及脚踝或脚履。从外形上看，肖斯（Chausses）很像今天的长筒袜。伴随着波尔普万（Pourpoint）和肖斯（Chausses）组合形式的出现，男裤女裙的服饰性别差异也开始形成。

14世纪中叶，出现了男女衣服造型上的分化，与男子服短上衣和紧身裤组合这种上重下轻的、富有机能性的造型相对，女服上半身紧身合体，下半身裙子宽大、拖裾，上轻下重，更富装饰性。这些与古代的宽衣文化性质完全不同，而是日耳曼窄衣文化的发展。

性别遮蔽

14 世纪末期，西欧最后一款男女通用式样的筒形衣吾普朗多（Houppelande）开始流行。最初，这种衣服是作为对当时变短的男服的一种弥补而出现的室内衣，不久被女子采用，并逐渐发展为室外穿的盛装（图 6–45）。该款式也在以男性为主角的统治阶级和大商人中流行（图 6–46）。无论男女，吾普朗多（Houppelande）的最大特征是不显露体形，只注重衣服外表装饰，这就与同时期的科塔尔迪（Cote Hardi）和波尔普万形成强烈的对比，即一方是对肉体的肯定，一方是对肉体的否定，反映出中世纪西欧人在神权统治下，在禁欲主义支配下扭曲的矛盾心态。15 世纪晚期开始，筒形衣完全退出了男装的主流发展，成为女装的专用形式。

哥特式样

哥特式建筑的外观特征是采用锐角的塔和尖顶拱券。这一时期的服装受到了这种建筑风格的影响，出现了名为汉宁（Hennin）的尖顶高帽和尖头鞋（图 6–47 ～图 6–49），以及大下摆的裙子。这些与哥特式建筑向上升腾的垂直线特征一脉相承。这类服饰的出现使哥特式服装与哥特式建筑从神似达到了形似。

与此同时，这一时期的织物与服装所具有的光泽和鲜明色调也与哥特式教堂内彩色玻璃有着异曲同工之处。

图 6–46　男式 Houppelande 人物形象

图 6–47　具有哥特式建筑外观特征的尖顶帽

图 6–49　哥特风格深 V 领下摆拖地的女袍

图 6–45　绘画中身穿 Houppelande 的女性形象

图 6–48　具有哥特式建筑外观特征的尖头鞋

第七章 公元 14—17 世纪

第一节 中国：明代

明代：1368—1644 年

元朝末年爆发的农民起义，动摇了元朝的统治。明洪武元年（1368 年），朱元璋凭靠农民起义的强势力量，推翻了元朝统治，建立起明朝帝国，定都应天（今南京）。朱元璋在建立了明中央集权的封建帝国后，废除了有近千年历史的丞相制度和有七百多年历史的三省（中书、门下、尚书）制度，由皇帝一人统揽大权。明王朝共历 276 年统治，将国号定为“明”，寓意“光明”。

明代疆域最盛时，北达乌第河，东达日本海，西达哈密。15 世纪初，郑和七次下西洋成为中国乃至全世界航海史上的伟大壮举。明代的文化也被后人珍藏，特别是《水浒传》《三国演义》《西游记》等小说作品闻名于世。宋应星、徐光启等科学家的贡献名不虚传。明代还出现了中国历史上规模最大的类书——《永乐大典》。

衣冠唐制

明太祖建立大明帝国后，对宋之后中国政权长期处于少数民族统治之下形成的习俗、礼仪制度进行调整改制，服饰礼仪更在改制之列。先是洪武元年（1368 年）废止了元代的服饰制度，继而又以明太祖的名义下诏：衣冠悉如唐代形制，用了近 30 年时间，恢复汉族礼仪与习俗，对皇帝冠服、文武百官服饰、内臣服饰、侍仪冠服、士庶冠服、乐工冠服等都做出了一一详尽的重新厘定。

明初服制规定臣民器服不得用黄，皇太子以下职官不置冕服。嘉靖七年（1528 年），明世宗为了

图 7-1 明世宗画像（南薰殿旧藏）

其式样为盘领大袖，窄袖口，领口右侧钉纽襻扣一对。大襟和小襟的外侧各钉罗带两根，左右腋下各钉罗带一根，用以系结固定。前襟和后背中部上、中、下各有三个团龙。

图 7-2 万历皇帝十二章圆领衮服（北京定陵出土）和结构图

达到在家“虽燕居，宜辨等威”的目的，接受大臣张璁的建议，以“燕弁”为名，寓“深宫独处，以燕安为戒”之意，推出了燕弁冠服。

皇帝服制

明代天子之服有六种，即衮冕服、通天冠服、皮弁服、武弁服、常服和燕弁服。

衮冕服，祭天地、宗庙、社稷、正旦、冬至、圣节、社稷、先农、册拜时穿用。衮服有十二章纹。玄衣八章，日、月、龙在肩，星辰、山在背，火、华虫、宗彝在袖，每袖各三。十二章衮服实例可见于南薰殿旧藏帝王像（图 7–1），在北京定陵万历皇帝墓出土的十二章圆领衮服正与前者对应（图 7–2）。

以龙纹（图 7–3）装饰的袍服自明代起被广泛应用在皇帝服装上，主要有四种式样（图 7–4）：第一种，即十二团龙十二章衮服；第二种，即四团龙袍，在前胸、后背和两肩各饰团龙纹一个，其中胸背为正龙，两肩为行龙；第三种，即柿蒂形龙袍，在盘领周围的两肩和胸背部形成柿蒂形装饰区；第四种，即过肩通袖龙襕袍，在第三种 b 式的基础上在袖部和前后襟下摆装饰有行龙的横襕（图 7–5、图 7–6）。

皇帝亲征遣将时穿武弁服（图 7–7）。首服弁上锐，色用赤，上十二缝，中缀五彩玉，落落如星状。身上穿韎衣、韎裳、韎韐，俱赤色。佩、绶、革带，如常制。佩绶及韎韐，俱上系于革带。舄如裳色。

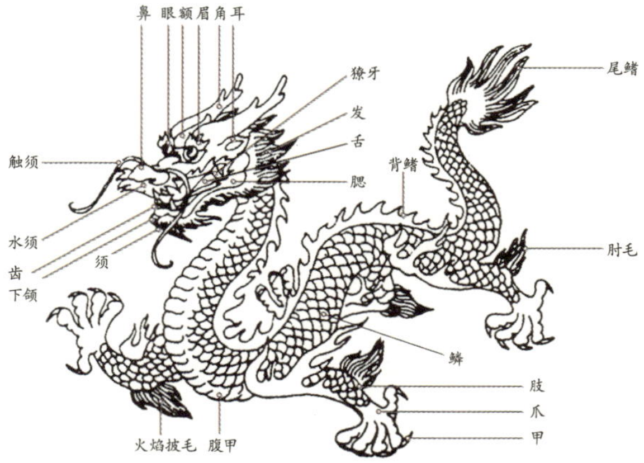

图 7–3　明代龙纹各部位名称

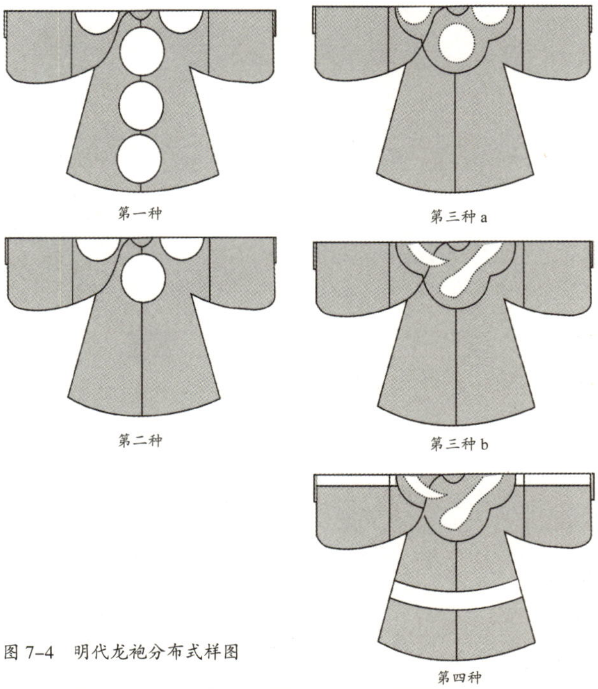

图 7–4　明代龙袍分布式样图

图 7–5　明太祖朱元璋像（南薰殿旧藏）

图 7–6　明宣宗朱瞻基像（台北故宫博物院藏）

图 7–7　《出警入跸图》中身穿戎服骑马出行的明代皇帝（台北故宫博物院藏）

朝服

明代官员在大祀庆成、正旦、冬至、圣节及颁降开读诏赦、进表、传制时穿朝服，头戴梁冠，以冠上梁数为差（图 7–8）。朝服为赤罗衣，白纱中单，青饰领缘，赤罗裳，青缘，赤罗蔽膝，大带赤、白二色绢，革带，佩绶，白袜黑履（图 7–9）。

公服

明代官员每日早晚朝奏事及侍班、谢恩、见辞时穿公服。在外文武官，每日公座服之。头戴幞头，展脚长一尺二寸，（杂职官的幞头，最初为垂带，后用展脚，与入流官同），身穿盘领袍，袖宽三尺（图 7–10、图 7–11）。

图 7–8　颖国武襄公杨洪像（美国国立亚洲艺术博物馆藏）

图 7–9　明代五梁冠（山东博物馆藏）

图 7–10　陆昶画像（南京博物馆藏）

图 7–11　明代《中东宫冠服》中的公服

图 7–12　明代身穿常服的沈度像

图 7–13　头戴幞头、身穿蟒服的明代官员王鏊像

图 7–14　明代麒麟吉服袍（山东曲阜孔府藏）

图 7–15　斗牛

图 7–16　飞鱼

常服

明官员凡常朝视事、日常生活时，头戴乌纱帽，身穿团领衫（图 7–12）。其因圆领袍胸前、背后各缀一方形补子，故也称补服。文官绣双禽示文明：一品仙鹤，二品锦鸡，三品孔雀，四品云雁，五品白鹇，六品鹭鸶，七品鸂鶒，八品黄鹂，九品鹌鹑。武官绣兽示威武：一品、二品狮子，三品、四品虎豹，五品熊罴，六品、七品彪，八品犀牛，九品海马。杂职：练鹊。风宪官：獬豸。

赐服

除补服之外，还有皇帝特恩授予特定人物的赐服（图 7–13），即一品斗牛，二品飞鱼，三品蟒，四品、五品麒麟，六品、七品虎、彪等纹样袍服。蟒是四爪的龙形动物。麒麟为传说中的鹿身、牛尾、马蹄、鱼鳞状的仁兽（图 7–14）；斗牛的外形与龙、蟒相类，斗牛角为弯曲状，龙的角为直角（图 7–15）。身上有翼和鱼尾的蟒形动物称飞鱼（图 7–16）。在明代，飞鱼服是仅次于蟒服的一种赐服。

曳撒

曳撒，其式当是前代“质孙”服的遗制，大襟右衽、长袖，后身为整片，前身为两截，腰下有细褶（中间不折），两腋缀摆。明初用于官吏及内侍（图 7-17）。有官者衣色用红，上缀补子；无官者衣色用青，无补。至明代晚期，演变为士大夫阶层常服，礼见宴会均可着之。

图 7-17　《朱瞻基行乐图》中身穿曳撒的人物形象

图 7-18　《朱瞻基行乐图》中身穿贴里的皇帝形象

贴里

贴里是明代内臣（宦官）所穿的袍服，大襟窄袖，下长过膝。膝下施一横襕，所用颜色有所定制，视职司而别（图 7-18、图 7-19）。明初规定，近侍用红色，缀本等补子；其余宦官用青色，不缀补子。万历年间（1573—1620 年），宦官魏忠贤专权揽政，于蟒贴里膝襕下再加一襕，称三襕贴里，两袖上各加两条蟒襕，另在胸、背等处织绣各式图纹，遍赏于近人亲信。

图 7-19　柿蒂窠过肩蟒妆花罗贴里（北京南苑苇子坑明墓出土）

图 7-20　清人徐璋绘《松江邦彦画像册》中的南太仆寺卿林景旸像

图 7-21　明衍圣公一品蓝暗花纱缀绣仙鹤交领直身（山东曲阜孔府旧藏）

直身

直身为道士、僧人所着之袍，以素布为之，交领右衽，大袖宽摆（图 7-20、图 7-21），也有在衣缘四周镶以黑边。明代时为士庶男子闲居、礼见所穿的服装。明代皇帝直身色用大红，普通人只准用天青、黑绿、玄青等色。直身两侧开裾，大、小襟及后襟两侧各接一片摆在外（共四片），有些会在双摆内再各加两片衬摆。双摆的结构是区分道袍和直身的标志。

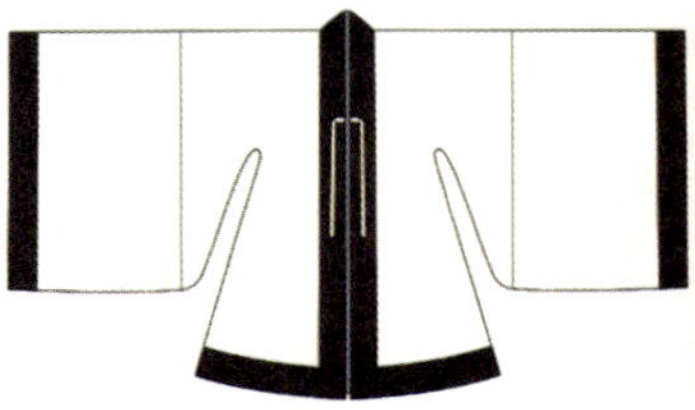

图 7-22　明代鹤氅示意图（贾玺增绘）

图 7-23　明代士人像

鹤氅

鹤氅是中国古代隐士、仙人、道士或文人雅士所穿服装（图7-22、图7-23）。鹤氅在晋代已有记载，如《晋书·谢万传》载：“著白纶巾，鹤氅裘。”因为宋代统治阶层笃信道教，从而导致文人雅士慕道之风盛行。到了明代，除了传统深灰色和青色，还流行浅色衣身的鹤氅，领袖衣襟均施有深色缘边。

凤冠翟衣

明代皇后在受册、谒庙、朝会时头戴九龙四凤冠（图 7–24），身穿袆衣，内套玉色纱中单。佩饰为玉革带。五彩大绶一，小绶三。玉佩两副。青色描金云龙袜、舄，每舄首饰珠五颗。北京定陵出土的明代凤冠中有“十二龙九凤冠”（图 7–25）。

黄色大衫

皇后常服为首服翠叶凤冠，身穿黄色大衫（图 7–26、图 7–27）。深青色绣团龙四襈袄子（即褙子）。红色鞠衣，配饰为红线罗大带，黄色织金彩色云龙纹带、玉花彩结绶。青袜舄。

图 7–24　《明代孝和皇后像》（南熏殿旧藏）

图 7–25　明代十二龙九凤冠（北京定陵出土）

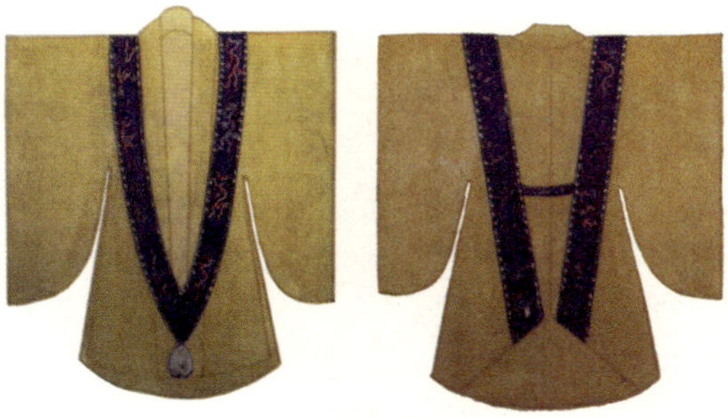

图 7–26　明代《中东宫冠服》中的黄色大衫

图 7–27　大衫实物（江西南昌华东交通大学校园内明代宁靖王夫人吴氏墓出土）

图 7–28　明代赭红凤补女袍（山东曲阜孔府旧藏）

长245厘米，宽13厘米，罗质，一端裁成斜边，后缝合（呈尖角），另缝三条横襻，以悬挂帔坠。两带以扣襻为中心分为前后两段，前段绣有四只翟鸟，后段绣三只翟鸟。每只翟鸟长约17厘米，展翅，直尾。

图 7–29　明代霞帔（江西南昌华东交通大学校园内明代宁靖王夫人吴氏墓出土）

图 7–30　明代凤纹金帔坠（江西南昌明墓出土）

鞠衣（皇后常服）

鞠衣为明代品官命妇冠服（图 7–28）。首服为特髻，其上按等级装饰翠松、口衔珠结金翟、珠翠花、珠翠梳、金云头连三钗、珠帘梳等饰物。服装为真红大袖衫、红色鞠衣（前后织金云龙纹，或绣或铺翠圈金，饰以珠）、红线罗大带、深青色霞帔、褙子等。

霞帔

明代霞帔（图 7–29）不仅应用更广，等级规定也更为详尽。由于礼服的穿用场合很少，于是在一些吉服场合（新婚、寿诞、画像留影等），霞帔脱离了大衫单独搭配圆领袍或云肩通袖袍。明代霞帔坠子制作也越发精美（图 7–30）。

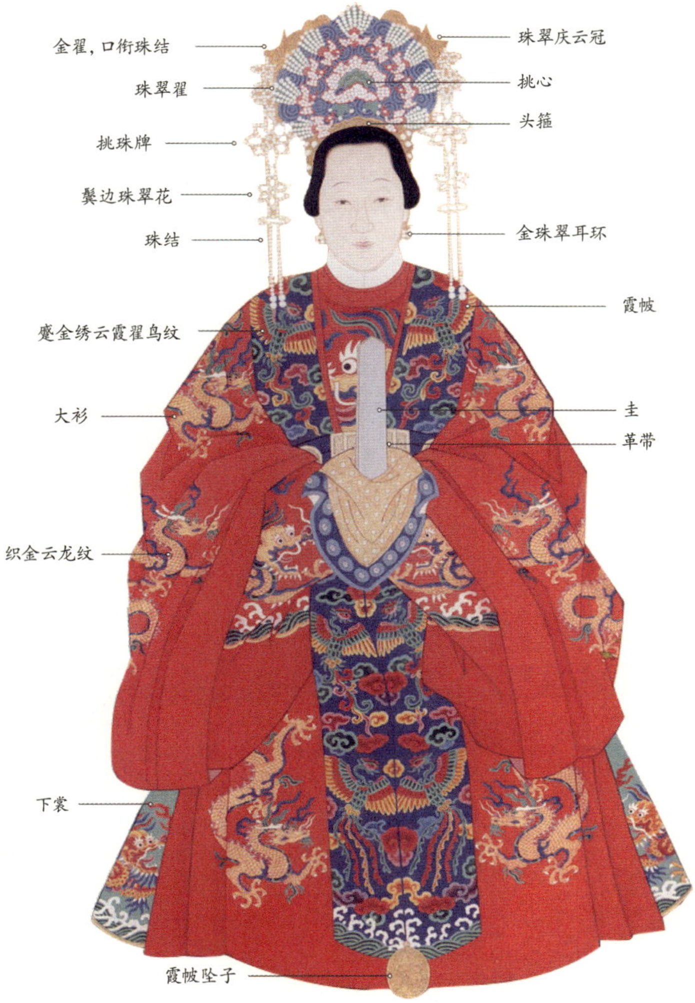

图 7-31　明代命妇礼服各部位名称示意图

图 7-32　明中期驼色暗花缎织金鹿纹方补斜襟短棉袄

图 7-33　明代驼色缠枝莲地凤襕妆花缎裙

图 7-34　明代印金膝襕女裙

襦裙

明代妇女常服主要有衫、袄、褙子、比甲及裙子等。明代女装大体分礼服（图 7–31）和常服两种。衣服的基本样式，大多仿自唐宋。上襦下裙的服装形式，在明代妇女服饰中仍占一定比例。上襦为交领、长袖短衣，有的还在其胸背部织绣有精美纹样的补子。其式样如北京地区出土的明中期驼色暗花缎织金鹿纹方补斜襟短绵袄（图 7–32）。

明代女裙式样更新，异彩纷呈：从质料上分有绫裙、绵裙、罗裙、绢裙、绸裙、丝裙、纱裙、布裙、麻裙、葛裙等；从工艺上分有画裙、插绣裙、堆纱裙、蹙金裙、细褶裙、合欢裙、凤尾裙等；从色泽上分有茜裙、郁金裙、绿裙、桃裙、紫裙、间色裙、月华裙、青裙、蓝裙、素白裙等，除了诏令严禁的明黄、鸦青和朱红外，裙色尽可随意。

裙子的颜色，在明代初期尚浅淡，虽有纹饰，但并不明显，如驼色缠枝莲地凤襕妆花缎裙（图 7–33）。至崇祯初年（1628 年），裙子多为素白，刺绣纹样仅在裙幅下边一两寸部位缀以一条花边，作为压脚（图 7–34）。裙幅初为六幅，即所谓“裙拖六幅湘江水”；后用八幅，腰间有很多细褶，行动则如水纹。

水田衣

水田衣是一种以各色零碎锦料拼合缝制成的服装，因整件服装面料色彩互相交错、形如水田而得名。水田衣具有其他服饰所无法具备的特殊效果，简单而别致，所以在明清妇女中间赢得普遍喜爱。据说在唐代就有人用这种方法拼制衣服，王维诗中就有“裁衣学水田”的描述。水田衣的制作，在开始时还比较注意匀称，各种锦缎料都事先裁成长方形，然后再有规律地编排缝制成衣。到了后来就不再那样拘泥，织锦料子大小不一，参差不齐，形状也各不相同，与戏台上的“百衲衣”十分相似，如明清江南画家绘《燕寝怡情》中身穿水田衣的女性形象（图 7-35）。

图 7-35　《燕寝怡情》中身穿水田衣的女性形象

百子衣

《红楼梦》第五十一回中有袭人身穿“桃红百子缂丝银鼠袄子”的记载。百子纹样为吉祥图案，寓意多子多福。宋代词人辛弃疾《稼轩词·鹧鸪天·祝良显家牡丹一本百朵》一词云：“恰如翠幙高堂上，来看红衫百子图。”北京明定陵出土的孝靖皇后的两件洒线绣百子女夹衣（图 7-36），衣上精致地绣有 100 个童子，共约 40 余个场景，有斗蟋蟀、戏金鱼、练武、摔跤、踢毽子、爬树摘果、站凳采桃、放风筝、玩陀螺、放爆竹、捉迷藏等。

图 7-36　孝靖皇后洒线绣百子女夹衣（北京明定陵出土）

图 7-37　红缎地盘金打籽绣云肩

云肩

明朝初期，云肩多为宫廷庆祝元旦及朝会的礼乐仪队中的女性歌舞伎所穿着（图 7-37）。随后，云肩普及到社会各个阶层女性，特别是青年女性婚嫁时不可或缺的衣饰。大多数云肩由四个云纹组成，称为四合如意式，还有柳叶式、荷花式等，上面都有吉祥纹样，如富贵牡丹、多福多寿、“连年有鱼”等。

第二节　西方：文艺复兴

14 至 17 世纪，西欧国家发生了伟大的资产阶级文化运动——文艺复兴运动（Renaissance）。中世纪的十字军东侵，打开了东西方的商路，意大利的佛罗伦萨、威尼斯、热那亚成了当时的贸易中心。资本主义的发展、新兴资产阶级财富的增加，使社会风气大为改变（图 7–38）。许多知识分子以古希腊、古罗马的古典文化为武器，向封建意识形态展开斗争。神的权威被人性所替代，人的价值和尊严得到肯定。教会的蒙昧主义和神秘主义被尊重知识、崇尚理性所替代。

这个时期造就了一大批对当时和后世都有很大影响的思想家、科学家和艺术家，诗人但丁（图 7–39）的《神曲》揭开了文艺复兴的序幕，达·芬奇（图 7–40）、米开朗基罗（图 7–41）、拉斐尔（图 7–42）和莎士比亚（图 7–43）等天才艺术家都为世人留下了不朽之作。

文艺复兴时期的服饰最显著的特点是服装由若干外形明确的独立部件组成。受欧洲各国国力消长和文化重心移动等因素的影响，文艺复兴时期的服装文化可大体分为三个阶段：意大利风时代、德意志风时代、西班牙风时代。

图 7–40　达·芬奇

图 7–41　米开朗基罗

图 7–42　拉斐尔

图 7–43　莎士比亚

图 7–38　文艺复兴时期纵情享乐的人们

图 7–39　诗人但丁

（一）意大利风时代（1450—1510 年）

意大利威尼斯、佛罗伦萨、米兰等城市都有高度发达的纺织作坊，可以大量生产天鹅绒、织锦缎和织金锦等华贵面料。为了尽量展开这些精美面料，服装出现了宽大平坦的平面造型。15 世纪中叶延续细长造型的男女装，到 16 世纪开始向横宽方向发展，男装变得雄大，女装变得浑圆。

女性造型重心放在下半身，头饰小巧，一般都不戴帽子，头发向后梳，颈后挽髻，露出额头。女服是在腰部有接缝的连衣裙罗布（Robe），领口开得很大，呈 V 型，也有一字型，袒露的胸口上装饰着珍珠项链，高腰身，衣长及地，袖子一段一段扎起来有如莲藕，在上臂、前臂和肘部有许多裂口（图 7-44）。内衣外露既丰富了服装配色，也方便人体运动。此时，袖子开始独立剪裁和制作，可以在服装上自由摘卸。为了收腰，女裙以腰围线为界，上下分别裁制，然后再缝合或用带子连接在一起。女士们流行穿高底鞋乔品（Chopines，图 7-45）。

在文艺复兴意大利风时代，男装重心放在上身，采用宽大的上衣波尔普万（Pourpoint，图 7-46）和紧身的下衣肖斯（Chausses，图 7-47）组合。波尔普万（Pourpoint）前开前合，用扣子固定和绗缝是其特点。自 14 世纪中叶至 17 世纪中叶，波尔普万（Pourpoint）作为男子的主要上衣延续了整整三个世纪。受西班牙服装式样影响，意大利风时代的波尔普万（Pourpoint）出现高立领，内衣的领子也随之变高，并且有褶饰（图 7-48）；衣长及臀，系腰带，前开式，用扣子闭合衣襟，腰部收细，袖子为紧身长袖，从肘到袖口用一排扣子固定。男子外出时一般穿齐膝大袍子和斗篷（图 7-49）。

图 7-44 文艺复兴风格女装

图 7-45 文艺复兴时期流行的高底鞋 Chopines

图 7-46 文艺复兴意大利风时期的上衣 Pourpoint

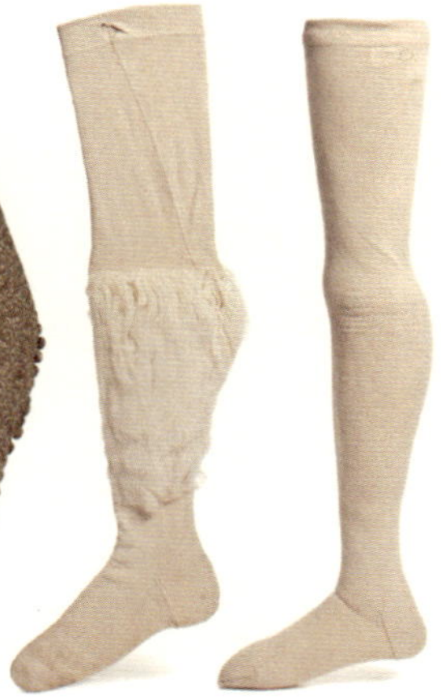

图 7-47 文艺复兴意大利风时期的 Chausses

图 7-48 立领有褶饰的内衣和 Pourpoint

图 7-49 外出时穿的大袍子和斗篷

（二）德意志风时代（1510—1550 年）

斯拉修（Slash）

德意志风时代服装的主要特色是露出里面的异色里子、白色内衣和缀饰有各色宝石、珍珠的裂口、剪口的斯拉修（Slash，图 7–50）装饰。它源自德国士兵刀剑的划痕，是男士们为了炫耀自己的英武而做出的一种人为装饰（图 7–51）。这种装饰也用在女装上（图 7–52、图 7–53）。

德国女服初期模仿意大利，为方形低领口，后来领口缩小，变成高领，袖子上有斯拉修（Slash）装饰（图 7–54）。女子服装上身腰很细，袖根部极其肥大，在细腰的下面收许多纵褶，使裙子在造型上形成鲜明的体积对比（图 7–55）。女子的大帽子上绣有花纹，饰有珠宝或鸵鸟羽毛，帽边有斯拉修（Slash）装饰。

图 7–50　身穿 Slash 装饰服装的萨克森公爵夫妻

图 7–51　《法国国王弗朗索瓦一世像》中服装袖子上有开口

图 7–52　袖子上有开口的文艺复兴意大利风时代的女装

图 7–53　Slash 装饰的女装

图 7–54　袖子上有 Slash 装饰的女装

图 7–55　文艺复兴德意志风时代的女服

茄肯（Jerkin）

这时带裙身的茄肯（Jerkin，图 7–56）取代了波尔普万（Pourpoint），它有很高的立领且有普利兹褶。男子在紧腿裤肖斯（Chausses）的外面，再增加鼓胀的短裤布里齐兹（Breeches）和阴茎套科多佩斯（Cod Piece，图 7–57）。这种装饰在欧洲非常流行。外出时，人们会在外面敞怀套衣身宽大的、袖子可拆卸的、毛皮里子或毛皮边饰的披肩。男子常在大帽里面再戴一顶软帽。

德意志风时期，人们还没有形成卫生观念，据说他们每年洗两次澡，因此，贵夫人们在脸上涂很厚的粉和胭脂，为了掩饰身体的异味，香水应运而生，各种化妆品制造业也蓬勃兴起。法国国王亨利三世时，男人也使用化妆品和香水，睡觉时同样戴面罩和手套。男女都时兴在头发上撒紫罗兰香粉。

图 7–56　德意志风时代的 Jerkin

图 7–57　奥地利蒂罗尔的斐迪南大公肖像中的 Cod Piece

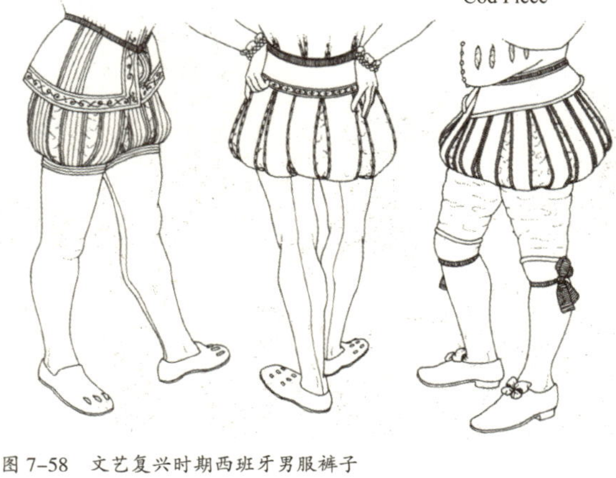

图 7–58　文艺复兴时期西班牙男服裤子

（三）西班牙风时代（1550—1620 年）

16 世纪是西班牙的世纪，西班牙拥有举世闻名的无敌舰队。西班牙国王强制性地向欧洲各国推行西班牙服装，企图使人们顺应西班牙意志。国力强大的西班牙在这一时期成为欧洲的流行中心。

填充物

西班牙服装的外观特征是大量使用填充物（图 7–58），在波尔普万（Pourpoint）的肩部、胸部、腹部和袖子都塞进填充物使之膨起，甚至短裤和阴茎套也使用了填充物（图 7–59），并装饰得格外引人注目。斯拉修（Slash）装饰也还有保留。外出时，他们在波尔普万（Pourpoint）外穿长及臀或膝的大翻领袍葛翁（Gown）和常有毛皮边饰的曼特尔（Mantle）。他们常在伸胳膊的地方装饰有假袖子。

西班牙风时代用料极为奢华，此时流行在黑色底上用金银线刺绣华美的纹样，天鹅绒、织金缎、织银缎备受宠爱，织物纹样也变得复杂化，拜占庭、萨拉森、波斯、中国风格的纹饰都出现在西欧贵族服饰中。除了复杂的纹样和精美的刺绣，贵族们甚至毫不吝惜地在服装上缀满宝石和珍珠。有时由于衣服被装饰得过于沉重，以致着装者步履维艰，出现连做弥撒时都无法下跪的窘境（图 7–60）。

图 7–59　亨利八世肖像

图 7–60　文艺复兴西班牙风时代带有垂饰的女服

紧身胸衣苛尔·佩凯（Corps Pique）

西班牙风时代，封建贵族的傲慢与奢靡心态作用到女装上。先前的袒胸式大领口被高达耳根的拉夫领（Ruff）替代，因为女性迷恋于高傲的贵族风度，从而使头部失去了活动的自由（图7-61、图7-62）。

此时，塞满了填充物的袖子变得有些僵硬，为了便于缝制，袖子与衣身分开缝制，每次穿时用细带连接。于是，女服也与男服一样，出现了掩饰袖根接缝的肩饰。袖子成为一种可以更换搭配不同服装的可变元素，甚至还成为一种赠送友人的礼品。此时，还有一种极为特殊的带有垂饰的式样，即外套的袖子从肩部打开，露出里面的面料，在肘部固定，袖子长长垂至地面（图7-63）。

西班牙风时期的女服盛行使用紧身胸衣苛尔·佩凯（Corps Pique，图7-64）和裙撑法勤盖尔（Farthingale）。此时的紧身胸衣已经具有了完备的形制，成为一种塑造女性胸腰部位立体造型的独立部件。

当时的英国女王伊丽莎白一世（1558—1603年在位）曾一度极力倡导束腰，影响了当时女性的时尚。文艺复兴时期的女服以细腰阔裙的装束美而著称。刚刚从上下一体、无腰身和性别差异服饰盛行的中世纪走出的女性们，为了突出身体的曲线美，通过紧身胸衣和裙撑的组合对比来完成细腰身的造型。

图7-61　文艺复兴西班牙风时代的男女服装

图7-62　《伊丽莎白一世肖像》

图7-63　文艺复兴西班牙风时代带有Ruff的女服

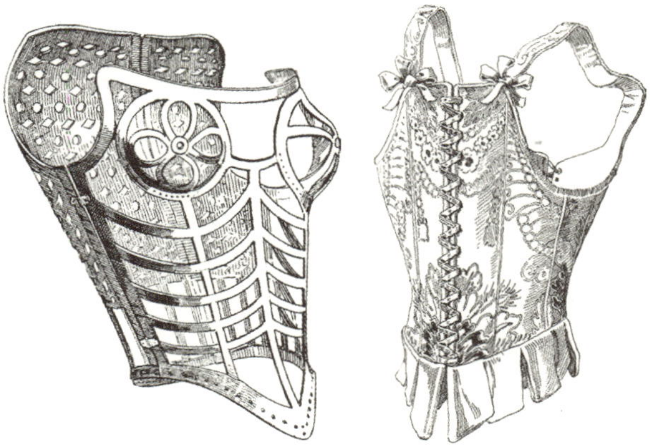

图7-64　文艺复兴西班牙风时代的紧身胸衣

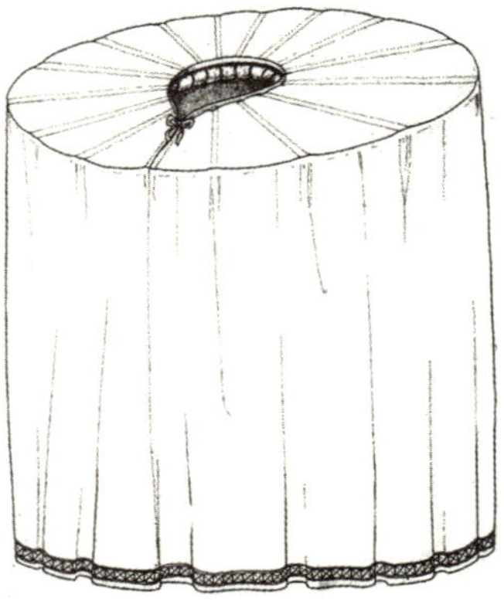

图 7-65　文艺复兴时期的裙撑 Wheel Farthingale

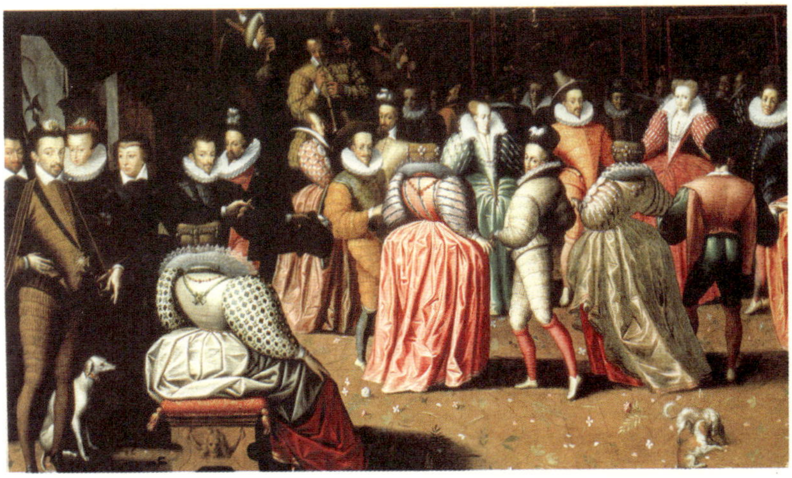

图 7-67　《宫廷舞会》

图 7-66　身穿西班牙风时代女装的贵族女性

图 7-68　西班牙裙撑（右图）和法国式裙撑（左图）

法勤盖尔（Farthingale）

西班牙的时尚一直影响着英国，即使在两国交战的情况下也依然如此。西班牙贵族发明的法勤盖尔（Farthingale），从 16 世纪下半叶到 17 世纪一直引领着欧洲时尚。这是一种在亚麻布上缝进鲸鱼轮骨或藤条的吊钟形裙撑。当时，不用做体力活的贵族女性都穿着西班牙法勤盖尔（Farthingale）。接着法国人又用马尾织物、铁丝和填充物做成了轮胎状的裙撑附加物威尔·法勤盖尔（Wheel Farthingale，图 7-65）。西班牙裙撑造型如吊钟形，附加了法式裙撑后使裙子从臀部看上去像个圆环，裙子的布料从女人腰间的撑环外缘垂直坠下（图 7-66、图 7-67）。两种式样同时存在于欧洲这个时期（图 7-68）。

罗布（Robe）

由于此时期前开式罗布（Robe）的流行（图 7-69），女性服装胸前连接处的装饰显得格外重要，这就产生了掩盖紧身胸衣苛尔·佩凯（Corps Pique）的装饰性三角胸衣斯塔玛克（Stomacker，图 7-70）。16 世纪中叶，为了强调女性的细腰，斯塔玛克（Stomacker）在前腰部接缝处呈锐角形下垂，并从其顶点向下呈 A 字型打开，露出里面的衬裙。因此，衬裙必须与外面的罗布（Robe）在用色和用料上保持协调，常使用厚地塔夫绸、织金锦等豪华面料，还装饰有宝石和精美的刺绣。

德意志风时代已见端倪的内衣高领领缘，到西班牙风时代，完全脱离内衣，成为一种独立制作、可以摘卸的拉夫领（Ruff，图 7-71 ～图 7-73）。可以说，拉夫领（Ruff）是欧洲 16 世纪到 17 世纪服装最显著的特征（图 7-74）。

图 7-69　文艺复兴西班牙风时代的前开式女裙

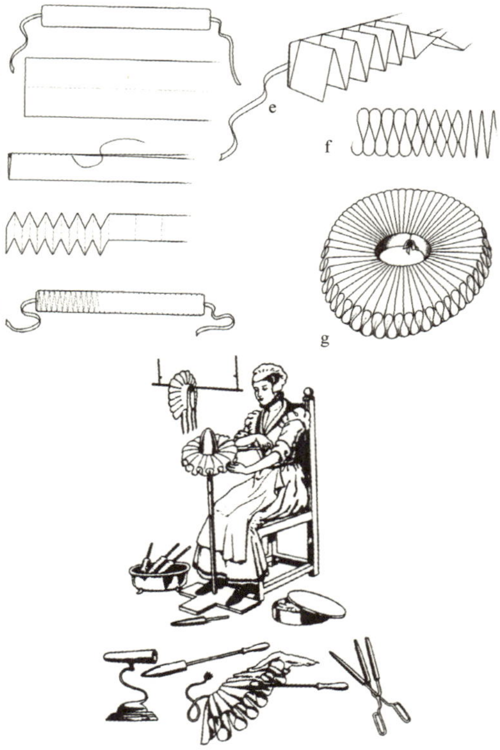

图 7-71　文艺复兴西班牙风时代 Ruff 的制作过程

图 7-70　装饰性胸布 Stomacker

图 7-73　文艺复兴西班牙风时代的男子

图 7-72　文艺复兴时期的 Ruff 漫画

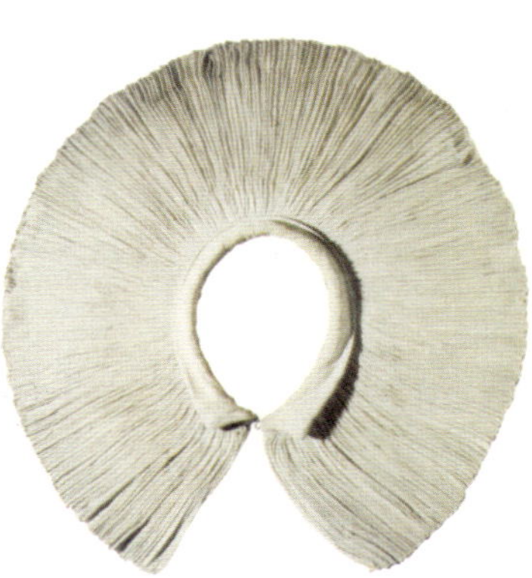

图 7-74　文艺复兴西班牙风时代的 Ruff 实物

第八章 公元 17—19 世纪

第一节 中国：清代

清代：1616—公元 1911 年

清朝（1616—1911 年）是中国以满族为主要统治阶级的最后一个封建王朝。1616 年，努尔哈赤（图 8-1）征服建州女真各部后建立了后金政权。1636 年改国号为清。1644 年，顺治帝福临入主中原。至 1911 年辛亥革命宣告清王朝灭亡，清王朝统治共计 295 年。

清政府统一多民族国家的疆域，加强内地与边疆的政治联系，密切各民族经济、文化交流。经顺治、康熙帝励精图治，清王朝统治者逐步扫清叛乱，采取有利于社会安定、经济发展的积极措施，社会生产得到发展，社会秩序日趋安定，由此出现了一个国家统一、政权巩固、社会安定、经济文化都比较繁荣的时期——“康（熙）乾（隆）盛世”。当时的手工业与商业取得了长足发展，手工艺术品制作十分发达，商品经济对民居的影响逐渐具体化。

图 8-1 清太祖努尔哈赤朝服像

在清之初，政府颁发了“剃发令”，强令其统治下的全国各民族改剃满族发型，实行“留头不留发，留发不留头”的高压政策。由于各族人民的强烈反对，在执行上变通为“十降十不降”制度。如“生降死不降”，指男子生前要穿满人衣装，死后可服明朝衣冠入葬；“老降少不降”，指儿童可穿明代或之前的传统服饰；“男降女不降”，指女子并未被要求改换服饰；“妓降优不降”，指娼妓穿着清廷要求穿着的衣服，演员扮演古人时则不受服饰限制；“官降民不降”，指官员在服饰上要求严格，但一般民间尤其是广大农村地区，仍可穿着明朝衣冠。这似乎是清朝初建时在服饰问题上与民间达成的一种“协议”，其实也是朝代交替变更期间服饰民俗演化的必然现象。

清代服饰制度，反映了清代社会政治制度的特点。它既借鉴汉族服饰中的某些元素，如以中国传统的十二章纹作为礼服和朝服的纹饰，以绣有禽兽的补子作为文武官员职别的标识，又不失其本民族的传统礼仪和服饰习俗。

朝服

清代以前，祭服和朝服不论君臣皆分制，并根据不同祭祀礼节的轻重，选择不同等级的祭服。到了清代，唯有皇帝才有祭服，其余人员朝、祭合一，这是清代冠服制度区别于其他朝代的地方。

清代皇帝朝服有朝冠（图 8–2）、朝服、朝珠（图 8–3）、朝带和套在朝服外的端罩（冬季之服）及衮服（百官穿补服）。

清皇帝冬朝冠上缀朱纬，帽檐顶饰金缕丝镂空金云龙嵌东珠三层宝顶，每层贯东珠各一颗，皆承以金龙四，余东珠如其数，上衔大珍珠一颗，在檐下左右各垂带，交系于颐下。清皇帝夏朝冠与冬朝冠形制相似，用织玉草或藤竹丝为之，外裱以罗。

清代皇帝朝服由披领和上衣下裳相连的袍裙组成，是清代皇帝登基、大婚、万寿圣节、元旦、冬至、祭天、祭地等重大典礼和祭祀活动时所穿的礼服（图 8–4）。其色以黄为主（图 8–5），南郊祈谷、常雩（求雨）用蓝，朝日用红，夕月用月色（浅蓝色）。

清代皇帝冬朝服一式（图 8–6），两肩和前胸、后背各绣正龙一条，列十二章纹，间以五色云下平水江崖。下裳襞积绣行龙六条间以五色云，下平水江崖。下裳其余部位和披领全裱以紫貂，马蹄袖端裱以薰貂，自十一月初一至正月十五穿着。清代皇帝冬朝服二式与一式的区别在于腰帷处有行龙五条。清代皇帝夏朝服（图 8–7）整衣形式和花纹皆与冬朝服之第二式相同，唯其袍边均沿片金缘，且据气温变化有缎、纱材质，以及单、夹之分。

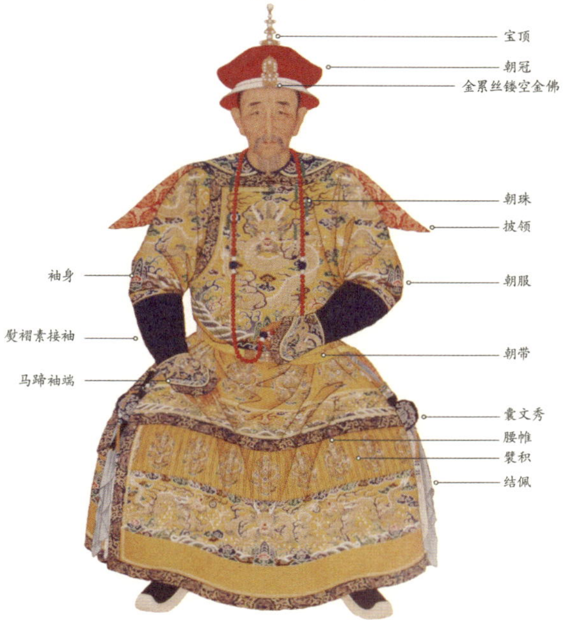

图 8–4　清代皇帝朝服名称示意图

图 8–5　清代皇帝朝服（蓝、黄、红和月色）

图 8–6　清代皇帝冬朝服一式

图 8–2　清代皇帝冬朝冠

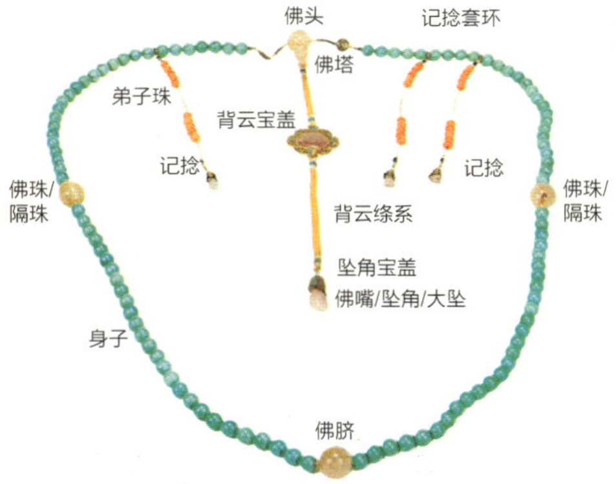

图 8–3　清代皇帝朝珠

图 8–7　清代皇帝夏朝服

衮服

衮服是清代皇帝祭天、祈谷、祈雨等场合套在朝服外面的外褂。其式样为圆领对襟，长与坐齐，袖与肘齐，石青色面，石青色扣鼻，五颗钻金圆纽子。其织、绣或缂丝五爪正面金龙四团为纹，前胸、后背、两肩各一，左肩日、右肩月，团龙间以五色云，下海水江牙。其实物如清宫旧藏乾隆石青色缎缉米珠绣四团云龙夹衮服（图 8–8），穿着效果如清朝皇帝溥仪衮服像（图 8–9）。

清代皇子所穿衮服只少日月纹，称为“龙褂”或“团龙补褂”，又因团龙补服上面所织的龙纹有四团：胸前一团，背后一团，左右肩各一团，故也称“四团龙补服”。团龙补服与皇帝衮服类似，是清代皇族才有资格穿的外褂。王公大臣和百官则无衮服，以补服替代。

身长 110.7 厘米，两袖通长 114.4 厘米，袖口宽 27 厘米，下摆宽 148 厘米。领口系残断黄条，黄纸签墨书『高宗』『缎绣缉米珠龙绵金』等。衮服石青色缎面料，用五彩丝线、金线和米珠在胸、背及两肩绣五爪正面龙四团，并在左右肩分别饰日月二章。在团龙纹样内，间饰五彩流云和红色万字、蝙蝠和寿桃纹，寓意『万福万寿』。龙纹均用小米般大小的白色珍珠绣成，装饰效果立体感强。

图 8–8 乾隆石青色缎缉米珠绣四团云龙夹衮服及局部图案

端罩

端罩，亦称“褡胡”，是一种皇帝、诸王、高级官员等人在冬季时替代衮服、补褂套穿在朝袍、吉服袍等袍服外的一种翻毛外褂（图 8–10、图 8–11）。其所用材料各有定制，以黑狐（亦称元狐或玄狐）为贵。其式似裘衣而较宽，长毛向外，圆领对襟，两袖手齐，下长过膝，左右衩微高，各缀一飘带。衬里则用软缎为之。文职四品及武职三品以下官吏，则无端罩。

下摆 125 厘米，开裾长 74 厘米，上半部为黑狐皮，毛长而具有光泽，下半部为貂皮，其毛尖均为白色，似一根根银针，是上等的貂皮料。其内衬明黄色暗花江绸里。端罩原有黄条，上书『黑狐皮端罩一件』。

图 8–10 黄色江绸黑狐皮端罩（清宫旧藏）

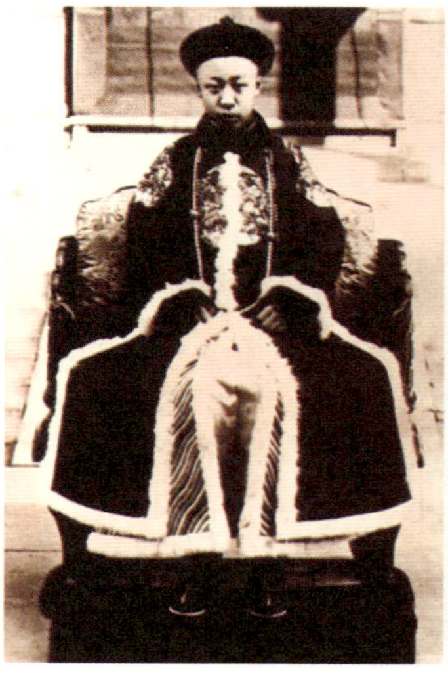

图 8–9 身穿衮服的清朝皇帝溥仪

图 8–11 外穿端罩的清代石文英像

图 8-12　清代皇帝绛丝十二章纹刺绣龙袍

图 8-13　《雍正帝读书像》

龙袍（吉服袍）

清代皇帝吉服也称龙袍（图 8-12），主要用于重大吉庆节日以及先农坛皇帝亲耕等场合，绣有九条龙（有一只被绣在衣襟里面）。《雍正帝读书像》中的雍正皇帝穿的就是明黄色缂丝刺绣龙袍（图 8-13）。从其正面或背面看，只能看到五条龙，九五之数象征帝王九五之尊的地位。除了龙纹，还有五彩祥云、蝙蝠、十二章纹等吉祥纹样（图 8-14）。另外，龙袍的下摆，斜向排列着许多弯曲的线条，名谓水脚。水脚之上，还有许多波浪翻滚的水浪，水浪之上，又立有山石宝物，俗称"海水江涯"。它除了表示绵延不断的吉祥含意之外，还有"一统山河"和"万世升平"的寓意。

与清代皇帝龙袍配套的服装有吉服冠（图 8-15）、吉服珠、吉服带（图 8-16）和配套的尖底靴（图 8-17）。

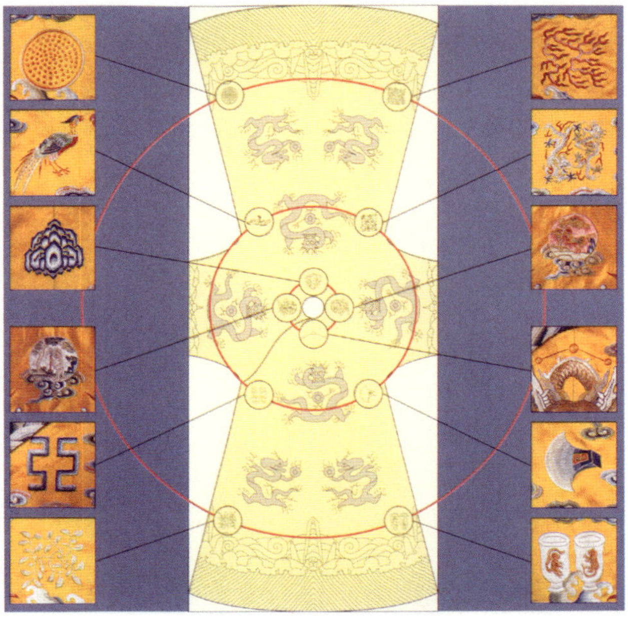

图 8-14　清代皇帝龙袍十二章纹分布图

图 8-15　清代皇帝吉服冠

明黄色丝织带，红色织金团龙缎衬里。带上饰镶珊瑚的方形带版四块，带版左右环上系湖色帨各一条，火镰一个、荷包四个、牙签筒一个、鞘刀一个。其中丝质挂件以金银线及五彩丝线施绣，绣工精细；金质挂件上镶绿松石。

图 8-16　清代皇帝吉服带（台北故宫博物院藏）

图 8-17　清光绪明黄色缎绣云蝠金龙纹尖底靴

后妃朝服

与皇帝朝服系统相对的是后妃们的朝袍（图 8-18）系统。清代皇太后、皇后、亲王和郡王福晋（满语“夫人”之意）及品官夫人等命妇的冠服与男服大体类似，只是冠饰略有不同。

清代皇太后、皇后朝冠，皆顶有三层，各贯一颗大东珠，各以金凤一只承接，每只金凤上饰东珠三颗、珍珠十七颗。皇太后、皇后、皇贵妃等人的冬朝袍皆为三式。第一式（图 8-19）：披领及袖皆用石青色，袍边和肩上下装饰片金缘（冬天饰貂缘），胸背部各饰正龙各一条，下摆前后各饰升龙两条（三条龙呈品字型），袖端饰正龙一条，披领饰行龙两条，两袖接袖（在综袖上，只有女袍有）饰行龙各两条，中无襞积。第二式：胸背柿蒂形纹样装饰区内前后饰正龙各一条，与第一式不同，下摆无八宝平水纹装饰。第三式：其裾左右后三开，其余皆与冬朝服第一式相同。

清代后妃朝服系统主要有朝冠（图 8-20）、耳饰（图 8-21，一耳戴三钳）、朝珠（图 8-22）、彩帨（图 8-23）、领约（图 8-24）和金约（图 8-25）。

图 8-18 清孝贤纯皇后朝服像

图 8-19 清代皇太后冬朝袍第一式

图 8-20 皇后夏朝冠

图 8-21 清代后妃耳饰

图 8-22 《皇朝礼器图式》中的清代皇后朝珠

图 8-23 清代湖蓝色琴棋书画纹样盘金绣彩帨

图 8-24 清代金点嵌珊瑚珠领约

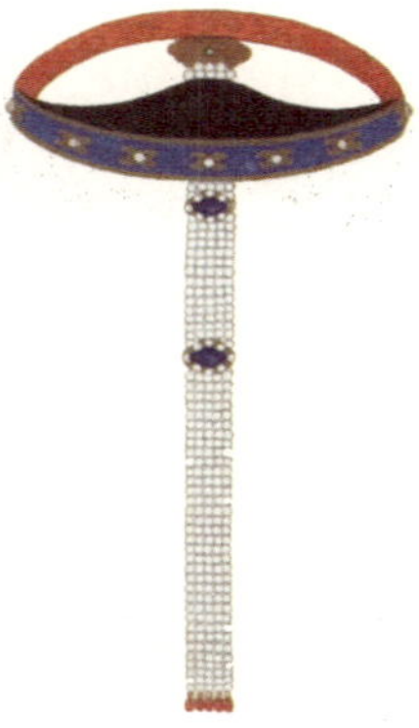

图 8-25 清代后妃金约

朝褂

清代皇太后、皇后朝褂可分三种式样，均为圆领、对襟、左右开衩、平袖，长与袍同，石青色，其主要特征为：第一式（图 8–26），在褂前后绣立龙各二；第二式，在褂上部织绣半圆形（内填织绣正龙两条，前后各一）纹饰；第三式，在褂前后织绣立龙各两条，中间没有襞积。

图 8–26　清代皇太后冬朝褂第一式和第三式

朝裙

朝裙是清代后妃至七品命妇于朝会、祭祀之时穿在朝袍里面的礼裙，均以缎为面料（图 8–27）。皇太后至三平命妇的朝裙有冬、夏之制。皇太后、皇后、皇贵妃朝裙款式为右衽背心与大摆斜褶裙相连的连衣裙（图 8–28），在腰线有襞积，后腰缀有系带两根，可以系扎腰部。贵妃、妃、嫔、皇子福晋朝裙膝以上用红缎。民公夫人、一品命妇朝裙，冬以片金加海龙缘，上用红缎面料，下用石青行蟒妆花缎面料；夏用缎或纱。

图 8–27　清代皇太后朝裙

图 8–28　清中期后妃冬朝裙

龙褂

清皇太后、皇后龙褂为圆领对襟，紧身平袖，左右开衩，长与袍相应的服装。皇太后、皇后龙褂有两种（图 8–29，图 8–30），其余人为一种。皇子福晋、亲王福晋、郡王福晋、固伦公主所穿就叫吉服褂而不叫龙褂。其着装效果如身穿龙褂的清代皇后像（图 8–31）。

图 8–29　清代皇后龙褂第一式
（清乾隆钦定《皇朝礼器图》）

图 8–30　清代皇后龙褂第二式实物

图 8–31　清代皇后身穿龙褂像

图 8-32 《曾国藩像》和身穿补服的清代官员画像

图 8-33 一品仙鹤暗花罗补服

图 8-34 清代霞帔

补服

清代普通官员穿补服，其形制为圆领对襟，平袖，袖与肘齐，衣长至膝，门襟有5颗纽扣（图8-32）。

补服服色无品级差别，乾隆时期至清末年间为石青色。清代补子比明代略有缩小，且只用单只立禽，前后成对，但前片一般是对开的，后片则整片织在一起。

亲王、郡王、贝勒、贝子等皇室成员用圆形补子；固伦额驸、镇国公、辅国公、和硕额驸、公、侯、伯、子、男以及各级品官，均用方形补子。文官补子：一品仙鹤（图8-33），二品锦鸡，三品孔雀，四品云雁，五品白鹇，六品鹭鸶，七品鸂鶒，八品鹌鹑，九品练鹊。武官补子：一品麒麟，二品狮，三品豹，四品虎，五品熊，六品彪，七品、八品犀牛，九品海马。

与明代单独穿用的圆领大袖补服不同，清代的补服是穿在吉服袍外。当皇帝穿衮服、皇子穿龙褂时，王公大臣和百官穿补服相衬配，故有“外褂”之称。根据清代制度，在每年盛夏三伏期间，可以不穿补服，被称作“免褂期”。

霞帔

明清两代命妇（一般为官吏之母、妻）亦备有霞帔，主要穿着于庆典朝会或吉庆场合。其形制为阔如背心的帔身，有后片和衣领，下施彩色流苏，在胸背正中缀与其丈夫官位相应的补子，显示身份（图8-34）。其纹样为：一、二品命妇霞帔纹样为蹙金绣云霞翟纹（翟，即长尾山雉）；三、四品命妇霞帔纹样为金绣云霞孔雀纹；五品命妇霞帔纹样为绣云霞鸳鸯纹；六、七品命妇霞帔纹样为绣云霞练鹊纹；八、九品命妇霞帔纹样为绣缠枝花纹。

武官之母、妻霞帔上的补子用鸟纹而不用兽纹，如武一品男补子织绣麒麟，武一品母、妻补子则织绣仙鹤，意思是女性娴淑，不宜尚武。

清代的平民妇女在出嫁之日可将霞帔作为礼服与凤冠一起穿戴。女补尺寸比男补略小，以示男尊女卑。

常服褂

常服褂即对襟平袖、长至膝下的外褂，皇帝及百官均穿着石青色，花纹不限，不缀补子，如《康熙帝读书像》中的服装式样（图 8-35）。

常服袍

常服袍为圆领右衽大襟，窄袖，马蹄袖端。常服褂套穿于常服袍外面。在清代皇帝、宗室所穿常服袍的腰部以下，都有开衩。因前后各一、长二尺余，左右各一、长一尺余，故名“四衩袍”，其形象如《玄烨便服写字像》（图 8-36）中的着装。官吏士人两开衩。

行袍

行袍，即清代文武官员出行时所穿长袍（图 8-37、图 8-38）。其式样为右衽大襟，窄袖，马蹄袖端，四开裾，右侧衣襟裁短一尺以便跨腿乘骑及开步射猎，故又称“缺襟袍”。凡臣工扈行、行围人员都可服之。文武官员出差时穿行袍谒客，外套对襟大袖马褂。

马褂

马褂为清代男子穿在长袍外，衣长及腰，袖长及肘，两侧及后中缝开衩的短褂（图 8-39）。衣襟式样主要有对襟、大襟和琵琶襟三种。袖口平直，无马蹄袖。女子马褂有挽袖（袖长于手）和舒袖（袖短于手）两种式样（图 8-40）。

清代马褂中等级最高和最具荣耀感的是黄马褂（图 8-41）。能穿黄马褂的有三种人：一是皇帝巡行扈从大臣可穿（正黄旗官员用金黄色），叫“职任褂子”；二是皇帝行围（狩猎）和阅兵时射箭或获猎物最多者可穿，叫“行围褂子”；三是治国、战功者可穿，叫“武功褂子”。前两种黄马褂用黑纽襻，平时不能穿用。第三种用黄纽襻，穿时随意。

清代黄马褂也有“赏给”与“赏穿”之分。“赏给”是只限赏赐之服，“赏穿”则可按时自做服用，不限于赏赐的一件。到清代中晚期，得此荣耀者为数较多。

图 8-35 《康熙帝读书像》

图 8-36 《玄烨便服写字像》

图 8-37 身穿缺襟袍射箭的清代人物照片

图 8-38 清代行袍（缺襟袍）

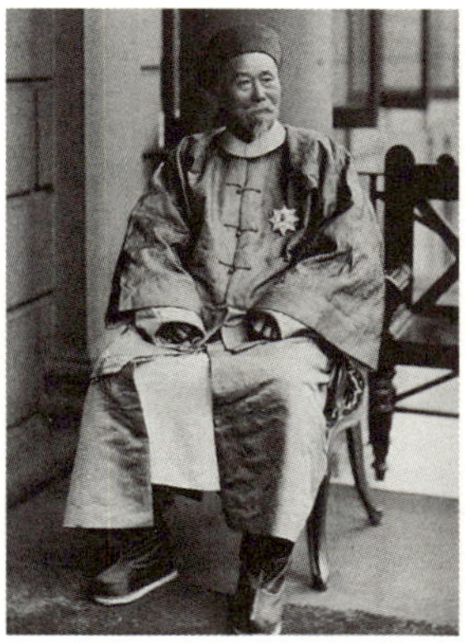

图 8-39 身穿马褂的清代官员李鸿章

图 8-40 身穿马褂的清代官员及命妇

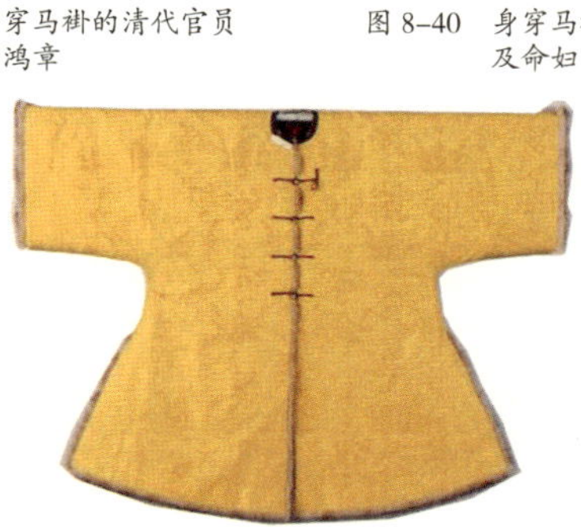

图 8-41 清代明黄缎棉马褂

女袍（吉服袍）

在时令节日等场合，清代后妃一般穿吉服袍。其式样为圆领大襟，右衽，左右开裾，直身，宽直袖。其实物如红缎绣仙鹤九团女夹袍（图 8–42）和红缎绣喜相逢九团女夹袍（图 8–43）。前者为圆领右衽，左右开裾，直身袍，大红缎地，绣九团仙鹤花鸟团花，以仙鹤和彩云组成。后者形制为圆领右衽，左右开裾，直身，大红缎地，绣九团“喜相逢”团窠图案。团窠直径 29.5 厘米，以蝴蝶上下盘旋相戏为主景，周围点缀蝙蝠及各种花卉。袖边平金织“卍”字曲水纹。蝴蝶图案采用 19 种颜色。刺绣针法有套针、戗针、缠针等。

图 8–42　红缎绣仙鹤九团女夹袍（首都博物馆藏）

图 8–43　红缎绣喜相逢九团女夹袍（首都博物馆藏）

图 8–44　月白色缎织暗花梅竹灵芝坎肩

图 8–47　清末月绿色绛丝云鹤纹夹坎肩

图 8–45　清代一字襟坎肩

图 8–48　宝蓝大提花葡萄纹琵琶襟坎肩

图 8–46　身穿坎肩的清宫女眷合影

坎肩

坎肩，也称马甲、背心，是无领、无袖的上衣，可套在长袍外起装饰作用（图 8–44）。清代坎肩用料、做工十分讲究，尤其是女性使用的坎肩（图 8–45），其衣襟式样有对襟、右襟和最具特色的“巴图鲁”（满语“勇士”的意思）一字襟坎肩（图 8–46），前襟上装有排扣，两边腋下也有纽扣。当时在京师八旗子弟中甚为流行。清代对襟坎肩实物如清末月绿色绛丝云鹤纹夹坎肩（图 8–47），身长 70 厘米、肩宽 40 厘米、下幅宽 80 厘米。前胸及开裾处饰如意云头五朵，通身镶青色云鹤纹绛金宽边及朵花绦。琵琶襟坎肩实物如宝蓝大提花葡萄纹琵琶襟坎肩（图 8–48）。

褂襕

清代还有一种无袖的褂襕，亦称背夹，冬季贵妇着旗装时穿在袍外。其实物如清光绪年间藏青缎地彩绣花蝶褂（图 8–49），长 148 厘米，立领，右衽，左右开裾。有库金、彩绣、中绦三道缘，褂面绣菊花、海棠花、兰花，彩蝶翩翩。

衬衣

满族贵族妇女便服流行长袍式衬衣（图 8–50）。衬衣多穿于氅衣或马褂内，也可单独穿用。日常穿着时多将袖口挽起，露出里面的白地缂丝袖口。领、袖和大襟均各镶两道边作装饰。衬衣既可单独穿着，也可以套穿于坎肩、马褂等短款或有开裾的长款褂襕、氅衣之内，形成面料花色的搭配或反差。

图 8–49　清光绪年间藏青缎地彩绣花蝶褂

图 8–50　清晚期明黄缎彩绣折枝玉兰花蝶衬衣

图 8–51　清晚期紫纱地缂丝百蝶散花纹氅衣

图 8–52　身穿氅衣的慈禧像

图 8–53　身穿旗袍的清代女子画像和照片

氅衣

氅衣是清代内廷后妃穿在衬衣外面的常服。其形制为直身，圆领，大襟右衽，左右开裾至腋下，顶端都有用绦带、绣边盘饰的如意云头，形成左右对称的形式，身长至掩足，只露出旗鞋的高底。双挽舒袖，袖端日常穿用时呈折叠状，袖长及肘，也可以拆下钉线穿用。袖口内加饰绣工精美的可替换袖头，方便拆换。衣边、袖端则装饰多种各色华美的绣边、绦边、滚边、狗牙边等，尤其是清代同治、光绪年间，这种繁缛的镶边装饰更是多达数层。其实物如清晚期紫纱地缂丝百蝶散花纹氅衣（图 8–51），紫色纱面，用三晕或四晕的缂丝织两两相对的蝴蝶和或单或双的折枝橘子花纹，时称“百蝶散花”，是慈禧（图 8–52）时期比较流行的纹样。清末，氅衣与满族服饰特点鲜明的袍服相融合，并逐渐演变成旗袍（图 8–53）。

马面裙

裙子在清代，不仅成了宫廷贵族小姐的日常装束，而且成了普通人家姑娘的日常服装。清代裙子的样式，初期尚保存明代特征，有凤尾裙及月华裙等。清末，普通妇女还流行穿裤，一为满裆裤，一为套裤。其材料多用绸缎，上绣各种花纹。

清代满族入关，虽然满族妇女都穿旗袍，但汉族妇女却仍把裙子作为礼服。每逢婚丧喜庆大典，或亲朋拜访时，妇女们一定要在长裤外面套一条长裙，否则就会被认为失礼或不够隆重。清代女裙，颜色以红为贵，丧夫寡居者只能穿黑裙。

马面裙（图 8–54）由两片相同的裙片组成，穿着时需要将裙腰上的扣子或绳系好。裙的两侧打褶裥，由两边向中间压褶，称“顺风褶”。裙腰多用白色布，取白头偕老之意。

清代前期，马面裙裙门图案复杂，有龙纹、凤纹、海水江崖、亭台楼阁、云纹和蝴蝶花卉图案等。这些图案在形式上呈现出适合纹样的特征，即图案底端适合裙门的四方形状，而上端则灵活多变。裙门的四周（腰头部分除外）皆有缘饰，从一条到多条不等。清末，裙门结构仍在，但是图案逐渐消失，最后仅剩缘饰。

马面裙两侧的褶如果打了又细又密的细裥，则称为“百褶裙”。为了使这些细褶不走形，以一定的方式用细线绗缝交叉串联固定褶裥，穿着者行走时，裥部形似鱼鳞鳞甲，故称“鱼鳞百褶裙”（图 8–55）。清末，马面裙的侧裥逐渐被“襕干”装饰物代替，因此这种马面裙又称为“襕干裙”。其形式与百褶裙相同，两侧打大褶，每褶裥镶襕干边（图 8–56）。裙门及下摆镶大边，色与襕干相同。

凤尾裙是一种由彩色条布接于腰部而成的条状女裙（图 8–57）。这些布条在末端裁成尖角，形似凤尾而得名。其中两条较阔，余均做成狭条。每条绣以不同花纹，两边镶滚金线，或缀以花边。背部则以彩条固定，上缀裙腰。因布条之间空隙较大，不能单独穿着，常作为附属服饰围系于马面裙之外（图 8–58）。装饰与马面裙有对应之处，前后有似裙门的平幅，平幅上也绣有龙凤图案。各布条上皆有绣花，有的还在凤尾下端缀有小铃铛。清代末年出现了将凤尾裙缝于马面裙之外的形制，两者合二为一，故称凤尾马面裙。

图 8–54　清代大红色三蓝绣花蝶图马面裙

图 8–55　银红暗花绸彩绣花鸟纹鱼鳞百褶裙

图 8–56　清代马面襕干裙

图 8–57　彩色缎地绣龙凤纹凤尾裙

图 8–58　清代五彩云纹暗花绸彩绣百褶凤尾裙

套裤

清代妇女在袍服或裙子里面穿直管状的膝裤（图8–59），也称“套裤”。这是一种裤腿上下垂直，呈直筒状，上端裁制成尖角状，和腰部相连，穿时可以露出臀部及大腿外侧的服装（图8–60）。清人李静山《增补都门杂咏》诗云：“英雄盖世古来稀，那像如今套裤肥？举鼎拔山何足论，居然粗腿有三围。”

除了套裤，清代女子还穿开裆裤（图8–61）。如红云纹暗花绸五彩绣花鸟纹女开裆裤，其裤腰为白色，象征白头偕老。穿着长裤时，多用色彩长汗巾在腰间系扎，垂露衣外，以为装饰。

清代套裤所用质料有缎、纱、绸、呢等，也有做成夹裤或在夹裤中蓄以絮棉的，后者多用于冬季。妇女所穿的套裤，裤管下脚常镶有花边，所用布帛色彩也较鲜艳。

图8–59 荷蝶绣暗花绸膝裤

图8–60 清代膝裤

图8–61 红云纹暗花绸五彩绣花鸟纹女开裆裤

图8–62 如意云纹黄朝靴

图8–63 詹天佑照片

图8–64 湖色缎绣兰花镶嵌宝石花马蹄底女鞋

图8–65 晚清紫缎钉绫凤戏牡丹花盆底女棉鞋

花盆底

清代男子便服穿鞋，公服穿靴（图8–62、图8–63）。靴多为尖头，黑缎制作。清制规定，只有官员穿朝服才许用方头靴。

满族妇女穿一种高木底绣花鞋，鞋底下小上大、似花盆的被称为“花盆底”（图8–64），鞋底下大上小的称为“马蹄底鞋”（图8–65），还有“龙鱼底”“四闪底”等，都是根据鞋子形状命名。鞋面多为缎制，绣有花样，鞋底涂白粉，富贵人家妇女还在鞋跟周围镶嵌宝石。老年妇女的旗鞋多以平木为底，称“平底鞋”。清代后期，着长袍穿花盆底鞋已成为宫中礼服。

清朝统治者入主中原后，起初极力反对汉人的缠足风俗，但缠足之风由来已久，康熙七年（1668年）只好罢禁。清代社会各阶层的女子，缠足之风日盛。

旗发扁方

汉族妇女的发髻首饰，清初大体沿用明代式样，普通女性简单地将头发绾至头顶盘髻。清代中后期，发式逐渐增多，有的模仿满族宫女发式，将头发分为两把，俗称“叉子头”；也有的在脑后垂下一绺头发，修成两个尖角，名“燕尾式”；后来还流行过圆髻、平髻、如意髻、苏州厥、巴巴头、连环髻、麻花等式样。此外，还有许多假髻，如蝴蝶、罗汉、双飞燕、八面观音等。满族未婚女子或蓄发在额前挽小抓髻，或梳一条辫子垂于脑后；普通已婚妇女多绾髻，有绾至头顶的大盘头、额前发髻鬅松的鬅头，还有架子头、在头顶左右横梳两个平髻的如意头（似如意横于脑后）、发髻梳成扁平状的一字头（图 8–66）。清代上层妇女一般留“两把头”。清末女性的发髻越梳越高，上套一顶形似扇形的冠，被称作“旗头”，即俗称“大拉翅”（图 8–67）。

满族妇女梳“两把头”或是“大拉翅”时插饰一种名为扁方的大簪（图 8–68、图 8–69）。扁方有玉制，也有用青素缎、青绒蒙裹而成，俗称“钿子”，佩戴时固定在发髻之上便可，上面还常绣有各种花纹图案，镶珠宝或插饰各种花朵且缀挂长长的缨穗。其作用是装饰并绾束发髻。在北方民间，扁方也有很小的。如遇丧事，妻子为丈夫戴孝，放下两把头，将头发集拢于头顶束起，分两把编成两条辫子，辫梢不系头绳，任头发松乱，头顶上插一个三寸或四寸长的白骨小扁方。如果儿媳为公婆戴孝，则要横插一个白银或白铜的小扁方。

图 8–66　梳“一字头”的清代满族妇女

图 8–67　梳“旗头”的清代婉容、文绣等人的合影

图 8–68　清代白玉荷叶莲花扁方

图 8–69　清代玳瑁镶珠石珊瑚松鼠葡萄扁方

图 8-70　绘画中的巴洛克时期剧院

图 8-71　巴洛克时期建筑室内装饰纹样

图 8-72　巴洛克时期建筑室内装饰

图 8-73　巴洛克风格绘画作品

第二节　西方：巴洛克、洛可可等

一、巴洛克

17 世纪的欧洲极为动荡，王权与各大小贵族、新兴资产阶级与封建君主势力、民众与资产阶级、代表旧宗教势力的天主教与代表革新派的新教之间，都展开了激烈的斗争。以德国为战场，几乎欧洲所有的国家都参加了这一举世闻名的 30 年战争（1618—1648 年）。

荷兰率先建立了第一个资本主义国家，英国经过一个世纪的反复斗争，也终于步入资本主义社会，法国则进一步强化了中央集权的专制政体。从 16 世纪以来就打下稳固的政治、经济基础的法国波旁王朝，在 17 世纪后半叶，由于“太阳王”路易十四推行绝对主义的中央集权制和重商的经济政策，法国国力得以发展，成为欧洲新的时装中心。从此，法国时装与法语、法国菜一起，成为人们追求的一种时髦。17 世纪的巴洛克风服装是以男性为中心，以路易十四的宫廷为舞台展开的奇特装束。

巴洛克一词本义是指一种形状不规则的珍珠，后指一种生机勃勃，有动态感，气氛紧张，注重光和光的效果，擅长表现各种强烈的感情色彩和无穷感的艺术风格的名称。在当时，巴洛克具有贬义，古典主义者认为它的华丽、炫耀的风格是对文艺复兴风格的贬低。但现在，人们已经公认，巴洛克是欧洲一种伟大的艺术风格（图 8-70 ～图 8-73）。

文艺复兴时期，人们看重男装的庄重感和体积感。到了巴洛克时期，社会上层男性们追求的是富有教养、风流举止和游戏般的情调，纤细和装饰性的女性式样成为男装的主导风格。因此，服装史学家习惯将巴洛克时期称作是“男装女性化”时代。

巴洛克时期的服装大体上可分为两个历史阶段：荷兰风时代和法国风时代。

图 8-74 《亨利利齐伯爵肖像》

图 8-75 巴洛克荷兰风时代的人物着装

图 8-76 荷兰风时代的男子服饰

图 8-77 荷兰风时代穿长裤的男子

图 8-78 荷兰风时代女装的大翻领

（一）荷兰风时代（1620—1650 年）

17 世纪前半期的荷兰风样式从 1600 年前后开始，逐步把西班牙风时代被分解的衣服部件组合起来，从僵硬向柔和、从锐角向钝角、从紧缚向宽松的方向变化。荷兰风时代的上衣肩部已不见文艺复兴时期的横宽和衬垫效果，取而代之的是自然舒适。男子夹衣的腰线上移并收腰，腰带多以饰带形式出现。胸和后背有开缝装饰，袖紧身，袖口饰有花边。男女服式抛弃了轮状皱领、填充物、紧身胸衣和裙撑，使服装变得舒适而柔和（图 8-74、图 8-75）。

"三 L"时代

荷兰风时代的三个特征是长发（Longlook）、蕾丝（Lace）、皮革（Leather），即"三 L"时代。

此时的波尔普万（Pourpoint）变长，盖住臀部，胸前和袖子仍有纵向的斯拉修（Slash），袖口翻折过来的克夫用白色蕾丝装饰。男子穿着的紧身裤被一种半截裤所取代，裤长及膝，并且在膝上用吊袜带或缎带扎口，装饰有蝴蝶结，这种半截裤叫做克尤罗特（Culotte，图 8-76）。后来到了 1640 年，出现了长裤（长及腿肚子的筒形裤），这是西洋服饰史上首次出现长裤（图 8-77）。贵族男士们为了追求骑士风度，常在右肩上斜挂佩刀，头戴装饰有羽毛的宽檐帽子，留着长而松散的头发、胡子和胡子尖，并披有甩在一个肩膀上的斗篷。

以新教思想为精神支柱的荷兰资产阶级，主张简朴美德，女服由僵硬感变得柔和自然，除了和男性服装一样的披肩领之外，还出现了大胆的袒胸样式，从西班牙的极端夸张的人工美造型恢复到那种胖乎乎的自然美的外形上来。虽然在荷兰风时代，拉夫领（Ruff）仍然在使用，但已经开始被大翻领拉巴（Rabat）逐渐取代（图 8-78）。领子分为缝合式和系带式，系带式是将领子用两条细带系在脖子上，让大领子披在肩上。人们在领子和袖口处大量使用蕾丝，这也是这个时期显著的特点。女性穿裙子都是穿两条或三条，分为里裙和外裙，面料轻薄，外出的时候会将外裙前面打开，故意露出里面的裙子。

图 8-79　17 世纪的 Justaucorpr

图 8-80　法国风时代的男子

图 8-81　17 世纪绘画中身穿 Justaucorpr 的人物形象及实物

（二）法国风时代（1650—1715 年）

17 世纪中叶，在波旁王朝专制下兴盛起来的法国取代了荷兰作为欧洲商业中心的地位。自路易十四开始，法国渐渐成为西方世界的中心。路易十四自称“太阳王”，喜欢用穷奢极欲来显示他的无上权威。受此氛围的影响，法国贵族阶层也沉迷于宫廷时尚和礼仪，无心政治纠葛和权术阴谋。

鸠斯特科尔（Justaucorpr）

法国风时代男装变化最明显，本来高腰线长衣摆的上衣波尔普万（Pourpoint）极度缩短，变成小立领，前开身，门襟上密密麻麻地缝缀着一排扣子，从右肩斜下的佩挂绶带表示身份（图 8-79）。1661 年出现了以缎带的使用量来显示身份高低的风气。此时还流行一种裙裤形式的半截裤（图 8-80）。

17 世纪 60 年代以后，男装再次出现市民性的贵族服，即由鸠斯特科尔（Justaucorpr）、贝斯特（Veste）和克尤罗特（Culotte）组成的男子套装。17 世纪 60 至 70 年代，鸠斯特科尔（Justaucorpr）被用作男子服，它衣身宽松，分有袖和无袖两种。后来这种服装逐渐从背缝和两侧缝收腰，并在两侧收褶、后背缝中下处开衩。收褶是为了使下摆张开，开衩是为了骑马时方便。17 世纪 80 年代，鸠斯特科尔（Justaucorpr）的腰身更加合体，形成了 19 世纪中叶以前的男服基本造型（图 8-81）。其口袋位置很低，整个造型重心向下移。袖子宽度也是越靠近袖口越大，袖口上还有一对翻折上来的袖克夫。其无领，前门襟密密麻麻排列着一排扣子，还装饰有金缠子丝绸纽。金缠子主要用于前门襟边饰和那一排排的扣眼和扣襻装饰。鸠斯特科尔（Justaucorpr）用料有天鹅绒或织锦缎，再加上金银线刺绣，十分豪华。虽然有许多扣子，但着装时一般都不扣。

由于鸠斯特科尔（Justaucorpr）衣襟是敞开的，所以领饰显得格外重要。在朗幕拉布流行期间，就曾使用取对裥的蕾丝做的领饰——克拉巴特（Cravate），这被认为是现在的领带的直接始祖。

鸠斯特科尔（Justaucorpr）里面穿的是贝斯特（Veste）。贝斯特（Veste）早期是长袖短身，后来为了配合鸠斯特科尔（Justaucorpr），其衣身变长，收了腰身，前门襟也有了很多扣子。再之后，贝斯特（Veste）演化成无袖的背心，衣长逐渐变短，是现代西式背心的始祖。

苛尔·巴莱耐（Corps Baleine）

巴洛克时期，女服仍喜好纤腰凸臀，依然采用紧身衣形态，而裙撑则被淘汰。女服上半身盛行使用紧身胸衣苛尔·巴莱耐（Corps Baleine，图 8–82）。它使得 16 世纪的上身细长造型重新流行。这一时期的紧身衣在腰部嵌入了许多鲸须，缝线从腰向胸呈放射状扇形张开。

此时女服的领口开得很大，袖子变短，露出小臂，整个袖子呈灯笼型。领子和袖口上装饰花瓣状多层蕾丝（图 8–83）。裙子至少要穿两条，外裙通常在腰围处有多褶，长度垂至地面。到了 17 世纪 80 年代，女服中出现臀垫，它使得女性臀部造型越来越膨大。在日常穿着时，上层社会女子一般把外裙卷起集中堆在后面，垂下来形成拖裾，衬裙显露出来，形成层层叠叠的美感。这是这种夸张后臀部的“巴斯尔”样式在西洋服装史上的第一次出现，直到 18 世纪至 19 世纪末，亦反复出现。

图 8–82 巴洛克法国风时代的紧身胸衣

图 8–83 巴洛克法国风时代的女装

二、洛可可

18 世纪西欧各国自然科学日渐发达，哲学家从过去假设上帝存在进而推论所有事物的工作，转换为依据实验和观察的理性方法去推论世间的万象；工业日益发达，民主思潮高涨，加之产业革命的发展与法国大革命的爆发，这些客观形势的转变，对于当时艺术的发展有非常大的影响。

巴洛克艺术尽管有呆板的礼仪，有形式上的骄矜和夸张，但它毕竟是一个阳刚的时期。紧随其后的洛可可艺术，大约自路易十四逝世（1715 年）时开始，风格更为讲究、矫饰、纤细和柔弱（图 8–84）。“洛可可”一词源自法国词汇“Rocaille”，其意思是指岩状的装饰，基本是一种强调 C 字型的旋涡状花纹及反曲线的装饰风格。这种风格流行于路易十五时代，又称“路易十五样式”。

图 8–84 蓬帕杜夫人画像

图 8-85 表现洛可可时期西方室内人物生活的场面

图 8-86 《爱的宣言》（1731 年）

图 8-87 洛可可时期的 Watteau Robe 实物

洛可可艺术风格表现在服饰上，是追求服装面料的质地柔软，花纹图案小巧，而且面料的色彩呈现明快淡雅和浓重柔和相并进的趋势。尽管一些欧洲国家屡次禁止印花棉布和丝绸的进口来保护本国纺织工业的发展，但因此而导致此类货物稀少，更助长了人们穿着的欲望，所以用印花棉布和丝绸做成的长袍、短衫一时成为最时髦的服装，为洛可可风格在服饰上的发展打下了坚实的基础。

进入 18 世纪的洛可可时期，巴洛克那种男性的力度被女性的纤细和优美所取代。洛可可女装放弃了西班牙钟式裙那种几何形状的严谨，保留了宽大的髋部和紧身的胸衣，有褶裥、荷叶边、随意的花边和隆起的衬裙，变得充满风情。

如果说巴洛克风尚是男人的世界，洛可可则是以沙龙为舞台展开的女性优雅样式。女性们作为供男性观赏和追求的"艺术品"和"宠物"，这种氛围使女装对外表形式美的追求发展到了登峰造极的地步，服装的每一个细节都精致化，最有品味的女性穿着"既暴露又优雅"（图 8-85）。

华托式罗布

洛可可风格在绘画方面最突出的代表画家是华托、布歇、弗拉戈纳尔和夏尔丹等人。从时间段来讲，华托是奥尔良公爵摄政时期的画家，布歇是路易十五与蓬帕杜夫人时期，即洛可可顶峰时期的画家。

让 - 安东尼·华托（Jean-Antoine Watteau，1684—1721 年）的绘画作品大多色调轻柔、形象妩媚，带有一种贵族阶层清高、典雅、朦胧的意境。华托在作品中经常描绘这种领口开得很大、背部有箱形褶的宽松袋状裙（图 8-86），所以这种裙子又被称为华托式罗布（Watteau Robe，图 8-87）或华托式袍（Watteau Gown），也有人将其称为麻袋背（Sack Back）。

由于这种优雅的样式主要为路易王朝宫廷及其周围的贵夫人穿用，特别是路易十五的情妇蓬帕杜侯爵夫人喜欢穿用，所以也称作法国式罗布（Robe à la francaise）。

图 8-88　洛可可女装中大量装饰鲜花

图 8-89　洛可可时期身穿 Pannier 的人物像

图 8-90　洛可可时期的女性形象

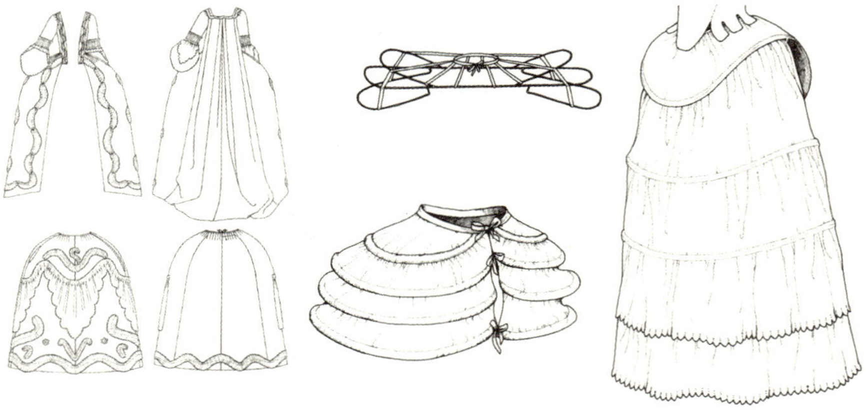

图 8-91　洛可可时期的 Pannier

花卉

洛可可服装的显著特点是柔媚细腻、纤弱柔和。其色彩常用白色、金色、粉红、粉绿、淡黄等娇嫩的颜色。自然形态在服饰上的体现就是以大量自然花卉为主题的染织面料。花朵在洛可可服饰上的运用除了表现在面料的图案上，还表现在大量运用天然或人造花朵对服装进行装饰。人们常把洛可可风格的女装比喻成“盛大的花篮”，在这个花篮里除了鲜花、蕾丝，还有蝴蝶结和缎带，使洛可可装饰艺术充满了女性惬意的轻松感，处处体现着新兴资产阶级上升阶段强调满足自身感官愉悦的审美趣味（图 8-88）。

荷叶边和褶边所形成的波浪形边缘轮廓线以及锯齿凸凹的外观效果，使衣边不再平直单调，而是层次起伏、轻盈飘逸，极具女性特征。

裙撑造型

在路易十五时代，一百多年前的裙撑又一次出现。初期裙撑帕尼艾（Pannier）为钟形，后来帕尼艾（Pannier）越变越大，逐渐变成前后扁平、左右加宽的椭圆形。

帕尼艾（Pannier）外层面积的增大给表层的装饰创造了更多的机会。前部敞开的罩裙以及裙子层次繁多是西方近代女装的重点。外裙下通常有内裙、衬裙和底裙。层层叠叠的裙子以它细腻精致、变化丰富的装饰形成层叠的视觉效果，成为 18 世纪人们追求矫饰和享乐的象征（图 8-89 ～图 8-91）。这种立体饰褶、服装面料的缝缀再造增加了服装在视觉上的浮雕感、立体感。1780 年，出现臀垫巴斯尔样式取代裙撑，而后臀部又一次膨胀起来，这种前腹稍平、后臀翘起的裙型称为巴斯尔样式。

图 8-92 波兰式罗布

图 8-93 托马斯·庚斯博罗《安德鲁夫妇》

图 8-94 18 世纪的男装

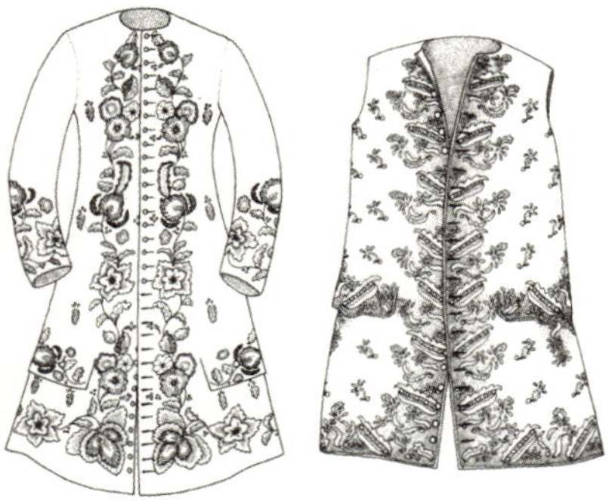

图 8-95 18 世纪的 Habit à la francaise 和 Veste

图 8-96 18 世纪的 Habit à la francaise 和 Veste

波兰式罗布

路易十六时代是洛可可风结束、新古典主义服饰样式兴起的转换期。18 世纪意大利那不勒斯两大古城的发掘，引起人们对古代文化的关注。人们开始从洛可可"优美轻薄"向"朴素、高尚、平静而伟大"的古典文化转移，此倾向被称为新古典主义。

1776 年，受波兰服装影响，开始出现波兰式罗布（Robe à la polonaise，图 8-92），裙子后侧像幕布或窗帘似的向上提起，为了把裙子束起，罗布的后腰内侧装着两条细绳，在表面同样的地方装饰着扣子或缎带，细绳从里面垂落，经裙摆向上把裙子捆束起来，绳端挂在或系在表面的扣子或缎带上，还有的在内侧裙摆处装上带环，绳穿过此环向上把裙子提起来后系上。此时，还流行英国式罗布，去掉巨大的裙撑，腰身下移，靠褶裥将裙子撑开，更加简洁、质朴，体现出英国的自然主义倾向（图 8-93）。

阿比（Habit à la francaise）

1760 年后，男性外套逐渐使用直线条，常伴随前短后长的设计，外套出现领子，袖子变得贴合手臂，背心长度变短，使用无袖及翻领（图 8-94）。

鸠斯特科尔（Justaucorpr）改称阿比（Habit à la francaise），收腰，下摆向外张，呈波浪状，臀部外张。为了使臀部外张，在衣摆里面加进马尾衬和硬麻布或鲸须。前门襟有一排以宝石等昂贵材质做成的扣子。

1715 年后，阿比（Habit à la francaise）大量使用浅色的缎子，门襟上的金缠子（扣子）装饰也省略了。由于阿比（Habit à la francaise）变得朴素，穿在里面的贝斯特（Veste）装饰得豪华起来，用料有织锦、丝绸及毛织物，上面有金线或金缠子的刺绣，衣长一般比阿比（Habit à la francaise）短两英寸（5 厘米左右，图 8-95、图 8-96）。下半身克尤罗特（Culotte）采用斜丝裁剪，做得十分紧身，不用系腰带，也不用吊裤带。其多用亮色的缎子，长度仍到膝部稍下一点，裤口用三四粒扣子固定。

18 世纪中叶，英国进入产业革命，男装也开始了新变革。男上衣去掉多余的量，衣摆不那么向外扩张，放松紧束的腰身，这种上衣称为夫若克（Frock）。其最大特点是门襟自腰围线起斜着向后下方剪裁。男士的鞋子上饰有"缎带""蔷薇花"造型并且男子普遍穿高跟鞋。高跟鞋最早起源于意大利，后来传到欧洲其他地方。

图 8-97 新古典主义时代西方女装形态变化

图 8-98 新古典主义时代的 Chemise 连衣裙

图 8-99 新古典主义时代的衬裙式连衣裙

图 8-100 新古典主义时代流行的披肩

三、新古典主义时代（1789—1825 年）

从女服样式变迁而言，研究者一般把 19 世纪女装分为以下五个时期（图 8-97）：新古典主义时期、浪漫主义时期、克里诺林时期、巴斯尔时期、S 型时期，直到 20 世纪 10 年代的霍布尔裙式样。因此，19 世纪也被称为“样式模仿的世纪”。

新古典主义兴起是以 1748 年庞贝城的发掘为契机，德国学者温克尔曼美学思想的传播引起了人们对古典主义的兴趣。古希腊、古罗马的英雄成了资产阶级所推崇的偶像，资产阶级革命家利用这些古代英雄号召人民大众为真理而献身。就在这样的历史环境下，产生了借用古代艺术形式和古代英雄主义题材大造资产阶级革命舆论的新古典主义。

修米兹（Chemise）

新古典主义时代的女装借鉴古希腊服装的特点，流行一种乳白色或浅黄色细棉布材质的，简练朴素的，领口有褶边的，方形或鸡心形袒胸的衬裙式修米兹（Chemise）连衣裙（图 8-98）。其造型特点是把腰线提高到乳房底下，用拉绳或腰带控制松紧，有很短的泡泡袖，将手臂裸露出来（图 8-99）。裙摆很长，柔和、优美的垂褶自高腰身处一直垂到地上。而且这种长裙越来越长，以至女士们行走时不得不用手提着裙子。这种优雅的姿态也成为流行时尚。

为了御寒，此时流行一种用开士米、薄毛织物、白色丝绸或薄地织金锦制成的披肩肖尔（Shawl，图 8-100）。它颜色鲜艳，围巾四角镶有细边和横贯围巾的长饰边，上面装饰着刺绣纹样。

图 8-101　新古典主义时代的 Spencer 短外套

图 8-102　英国式骑马装外衣

图 8-103　帝政时期有 Empire Puff 的女裙实物和画像

图 8-104　帝政时期分段扎成数个泡泡状的袖子

图 8-105　《约瑟芬皇后》

斯潘塞（Spencer）

与袒胸衬裙式修米兹（Chemise）连衣裙相搭配的服装是一种来自男服，叫作斯潘塞（Spencer）的短外套（图 8-101）。它类似西班牙斗牛士穿的短夹克，衣长仅达腰部，长袖。

此外，还有一种剪裁合体、有罩领的英国式骑马装外衣也是天气寒冷时的外衣。这种外衣门襟排列着一排扣子，装饰有金缠子丝绸纽（图 8-102）。

短帕夫袖（Puff）

到了帝政时期，修米兹（Chemise）流行白兰瓜形的短帕夫袖（Puff，图 8-103），也称帝政帕夫袖（Empire Puff），还有一种叫作玛姆留克（Mameluk）的袖子，是用细缎带把宽松的长袖分段扎成数个泡状的袖子（图 8-104）。短帕夫袖（Puff）主要用作仪礼服、宫廷宴会服；玛姆留克（Mameluk）长袖则主要用于外出服或家庭内穿的便服。

另外，当时流行两种颜色的裙子重叠穿用。外面的长罩裙会自腰围以下展开，露出里面修米兹（Chemise）的颜色（图 8-105）。

四、浪漫主义时代（1825—1850年）

浪漫主义是开始于18世纪的西欧的艺术、文学及文化运动，发生于1790年工业革命开始的前后。为了强调女性特征和教养，社交界的女士们经常随身带着药瓶，手里拿着手帕斯文地擦拭眼泪或文雅地遮在嘴上，故作纤弱、婀娜的娇态，好像久病未愈，弱不禁风。

浪漫主义时代的女装强调腰身与夸张裙摆。从1822年前后开始，女装腰线逐渐自高腰身位置下降，一直到1830年降到自然位置，腰又被紧身胸衣勒细，袖根部极度膨大化，裙摆向外扩张，裙子变为X型。1825年，裙摆变大，逐渐发展成吊钟状，1830年后，裙子的体积越发增大，衬裙数量达五六条之多。19世纪40年代初，还产生了用马鬃编成的钟形裙撑。

为了使腰显得细，女装肩部不断向横宽方向扩张，袖根部被极度夸张，甚至在袖根部使用了鲸须、金属丝做撑垫或用羽毛做填充物。低领的衣服多采用帕夫短袖（图8–106），高领的衣服多采用羊腿袖（图8–107）。这一时期的领子有两种极端形态：一种是高领口，上常有精饰，有时还采用16世纪的拉夫领（Ruff），也有像荷兰风时代的大披肩领一样重叠几层的有蕾丝边饰的披肩领（图8–108）；另一种是大胆的低领口，低领口上常加有很大的翻领或重叠数层的飞边、蕾丝边饰（图8–109）。1830年，还流行女子穿骑马服（图8–110）。

在19世纪20年代，浓丽的色彩与繁杂的装饰爱好再度展现在服装风尚里，华丽帽子上装饰着羽毛和蕾丝（图8–111）。社会上弥漫着乡愁的思绪，使得昔日传统富丽的装饰手法重现于扇子上（图8–112），以迎合此时维多利亚时代的繁华品味。

图8–106 19世纪20年代的Puff

图8–107 19世纪40年代的羊腿袖

图8–108 Ruff和大披肩领

图8–109 常加有很大的翻领或重叠数层的飞边、蕾丝边饰的低领口

图8–110 19世纪30年代的女子骑马服

图8–111 浪漫主义时代的女性帽子

图8–112 浪漫主义时代的女性扇子

五、克里诺林时代（1850—1870 年）

拿破仑三世执政的第二帝政时期，一方面复辟第一帝政的风习，一方面推崇路易十六时代的华丽样式的风格特征。理想的上流女子是纤弱并带点伤愁，面色白皙，小巧玲珑，文雅可爱，供男性欣赏的“洋娃娃”。这使女装再次回到洛可可趣味，故被称为“新洛可可时期”，又因女装上大量使用裙撑克里诺林（Crinoline），又称作克里诺林时代（图 8–113）。

图 8–113　Crinoline 裙撑

19 世纪 50 年代，紧身胸衣仍是不可缺少的整形用具，用来强调腰肢的纤细。同时，撑箍衬裙克里诺林（Crinoline）再度流行起来。最初的克里诺林（Crinoline）是用藤条或鲸骨制成的，裙撑支在平纹布衬裙的罩中。后来美国裙撑被普遍接受，这种裙撑由钟表发条钢外缠胶皮制成。撑箍衬裙在轻便的同时，也使女性的腰部更为苗条。撑箍衬裙庞大的体积常常使穿着者生活不便，即使是宽敞的舞厅和客厅也很容易变得拥挤不堪。当时欧洲关于克里诺林（Crinoline）裙撑的讽刺漫画有很多精彩的表现（图 8–114）。

图 8–114　克里诺林时代讽刺大裙撑的漫画

此时，女装的领子延续上个时代的高领口和低领口，高领口女装上有绣了花纹的领子，一般是前开襟，有一排或两排扣子。在女装上使用扣子来固定衣服就是从这个时代开始的（图 8–115）。低领口一般为四角形或 V 型的大开领，领口装饰着蕾丝。

图 8–115　克里诺林时代使用扣子的女装

由于裙撑的再次复活，罩在外面大裙子上的装饰也越来越多。19 世纪 50 年代，裙子表面横向布满了一段一段的襞褶装饰，通常分三段、四段、五段、七段不等（图 8–116）。极端者，如用奥甘迪（Organdie，蝉翼纱）做的裙子上有 25 段襞饰。在襞褶的设计上，有斯卡拉普（Scallops，海扇形缘饰）、流苏、缎带等装饰。这些装饰的色彩多采用裙子的对比色，十分鲜艳夺目。19 世纪 60 年代末，外出服出现了用四五个隐蔽的带子把外侧的罩裙卷起，露出里面衬裙的波兰风样式，这种样式只流行了两三年即转向后部突起的巴斯尔样式。

图 8–116　克里诺林时代女裙装饰的分段襞褶

六、巴斯尔时代（1870—1890 年）

1871 年，巴黎公社成立，时装界一度消沉，沃斯（Worth）的高级时装店关闭，裙撑被合体的连衣裙式的普林塞斯裙（Princess dress）取代。17 世纪末、18 世纪末两次出现过的臀垫巴斯尔（Bustle）又一次复活。因此，服饰史学家把 19 世纪 70 年代到 80 年代，将女性长裙裙身后曳部分堆放在后臀部的时期称为巴斯尔时代（图 8-117）。

当时女装流行巴斯尔样式，即撑架上蒙着马尾衬布，外侧的罩裙流行拖裾形式。女性服饰的臀部的夸张达到极限。巴斯尔时代的女装，除凸臀的外形特征外，另一个特色即拖裾。与后凸的臀部相呼应，这时女装在前面用紧身胸衣把胸高高托起，把腹部压平，强调“前挺后翘”的外形特征。这种极端的外形到 19 世纪 90 年代变为优美的 S 型。一般白天的领子为高领，夜间多为袒露的低领口。强调衣服表面的装饰效果是巴斯尔样式的又一大特征。因为女装上大量使用室内装饰手法，如窗帘的悬垂襞褶，床罩、沙发罩缘饰上所用的普利兹褶或活褶、荷叶边、流苏等，所以有人称其为“室内装饰业者”（图 8-118、图 8-119）。

1879 年夜礼服　　1870—1880 年小姐服　　1891 年舞会服

图 8-117　巴斯尔时代的女装式样

图 8-118　巴斯尔样式的女性着装

图 8-119　巴斯尔样式的帽子

图 8-120　19 世纪末女性广泛参与体育运动

图 8-121　S 型时代女性裙装

七、S 型时代（1890—1914 年）

从 1890 年起，巴斯尔（Bustle）从女装上消失，西方女装进入了 S 型时代。西方女装的紧身胸衣在前面把胸高高托起，压平腹部，勒细腰，在后面紧贴背部，把丰满的臀部自然地表现出来，从腰向下摆，裙子像小号似的自然张开，形成喇叭状波浪裙。从侧面观察时，人体挺胸收腹翘臀，宛如“S”型。服装史上将这一时期称为“S 型时代”（图 8-120、图 8-121）。在美国，因画家基布逊（Charles Duna Gilbson，1867—1944 年）喜画这种样式，故也称作基布逊外形（Gilbson girl silhouette）。该式样与新艺术运动所提倡的曲线造型一致。S 型时代的女装处在西方传统服装风格接近尾声，现代化女装时代即将来临的交汇点。

1906 年，巴黎时装设计大师保罗·波烈（Paul Poiret）推出高腰身的希腊风格，把数百年来束缚女体的紧身胸衣从女装上去掉，从此奠定了 20 世纪的流行基调，预示着腰身不再是女性魅力的唯一存在，这在服装史上是具有划时代意义的。1907 年，西方服装的 S 型设计逐渐趋缓，女装长度逐渐缩短，腰围放松，臀围收缩。S 型流行了 20 年左右，1908 年左右开始，女装向放松腰身的直线型转化，裙子也开始离开地面，露出鞋。S 型女装式样逐渐退出了历史舞台。

S 型女装顺应人体曲线造型，把服装设计的核心放到强调女性自身的优美体态上，是新艺术运动试图放弃传统装饰风格的参照，转向采用自然，如以植物、动物为中心的装饰风格的间接反映。新艺术运动最典型的纹样都是从自然草木中抽象出来的，多是流动的形态和蜿蜒交织的线条，充满了内在活力。而 S 型女装正是以人体的曲线美为强调重点，这是西方传统服装向现代服装转折的关键因素。

八、近代男装

从 1789 年法国大革命到 1914 年第一次世界大战爆发，这一个多世纪是西洋服装史的近代时期。这个时期的西方社会，无论是政治、经济还是各种文化现象都发生了激烈的变化（图 8–122）。随着法国君主制度的崩溃，特别是封建身份制度的崩溃，贵族男性们从宫廷舞会炫耀财富、沙龙里向女性献殷勤的事务中抽身，转向从事近代工业及商业等领域的务实性社会活动。男性们开始抛弃那些夸张且装饰过剩的服装，开始转向追求服装的品味性、合理性、活动性和机能性。可以说，与女装相比，男装首先迈入了走向现代服装的变革之路。

虽然法国大革命吹响了新时代号角，但是兴起于 18 世纪中叶的英国产业革命将男士服装的领导地位从法国转移到英国。机械的工业化大生产改变了资本主义经济和社会结构。各种科学文明的发达改变着人们的生活方式和生活意识。

图 8–122 《自由引导人民》

图 8–123 浪漫主义时期强调细腰的男装

代表庶民阶级的雅各宾派革命者在上衣里面穿双排扣的背心，下身穿长裤，头戴红色无檐帽。服装面料也由华美的丝织物变成朴素的毛织物。前襟自腰节开始向后斜着裁下去的夫若克（Frock），逐渐在男装中间普及。过去曾遭受鄙视的黑色，成为礼仪和公共场合的正式服色，具有新的权威。

从路易十八到查理十世的统治期间，法国上流社会的绅士们的生活充满了注重典雅和严谨的贵族风格。受同期浪漫主义女装的影响，男装也时兴收细腰身，耸起肩部（图 8–123）。男装的基本构成仍是夫若克（Frock）、长裤（Pantalon）和基莱（Gilet）的组合，夫若克（Frock）驳头翻折止于腰节处，前襟敞开不系扣，露出里面的基莱（Gilet），后面的燕尾有时长及膝窝，有时短缩至膝部稍上，肩、胸向外扩张，垫肩使肩部显得很宽，袖山处也膨鼓起来。与此相对，夫若克（Frock）强调细细的腰身，长裤也很细长，整体廓型呈倒三角形。男子外套鲁丹郭特（Redinggote）也同样是细腰身、下摆量加大的外扩型，旅行用大衣常装饰着披肩式短斗篷。

拉翁基茄克（Lounge Jacket）

19 世纪 40 年代中期的浪漫主义时期，干练的资产阶级实业家的装束成为流行风尚（图 8–124）。男装流行黑色和茶色。夫若克（Frock）的高领已变成像现在的西服领一样的翻驳领，长裤有宽裤腿和锥形裤两种。1848 年法国二月革命后，出现了今天西服上衣的前身——拉翁基茄克（Lounge Jacket，图 8–125）。有燕尾的夫若克（Frock）被作为礼服使用。

男装的基本样式仍是上衣、基莱（Gilet）和长裤（Pantalon）的组合。上衣有四种：白天的常服（Frock Coat），夜间正式礼服燕尾服（Tail Coat，也称作 Swallow Tailed Coat 或 Evening Dress Coat），白天穿的晨礼服（Morning Coat），外出便装拉翁基茄克（Lounge Jacket，意为休闲茄克）。有领座的翻领、袖口有硬浆的袖克夫的现代型衬衣和领带开始登场。

这时出现一种叫作“Inverness Cape”的有披肩的长袖大衣并作为男用夜间外套，腰部常系腰带。一般白天外出用大衣均无披风，一种带兜帽的大衣（Burnous）在男女中间流行（图 8–126）。19 世纪 50 年代出现了一种叫作“Raglan Coat”的插肩袖大衣。正式场合的帽子仍是大礼帽，平常则是毡帽、草帽或瓜形帽。

一直到第二次世界大战前为止，工人阶级和其他阶级在衣服上都有明显差别，但上层男装在式样上并无差别，只是面料和加工的精细程度不同而已。可以说这是在资本主义制度下，服装摆脱封建等级的束缚，向民主化方向迈进的一种进步（图 8–127）。而且这时男装（包括女装）开始分化出市井服（逛街服）、运动服和社交服等适合不同场合穿用的不同品种。

图 8–124　浪漫主义时期干练的资产阶级实业家的男子装束

图 8–125　1848 年二月革命时期的男装

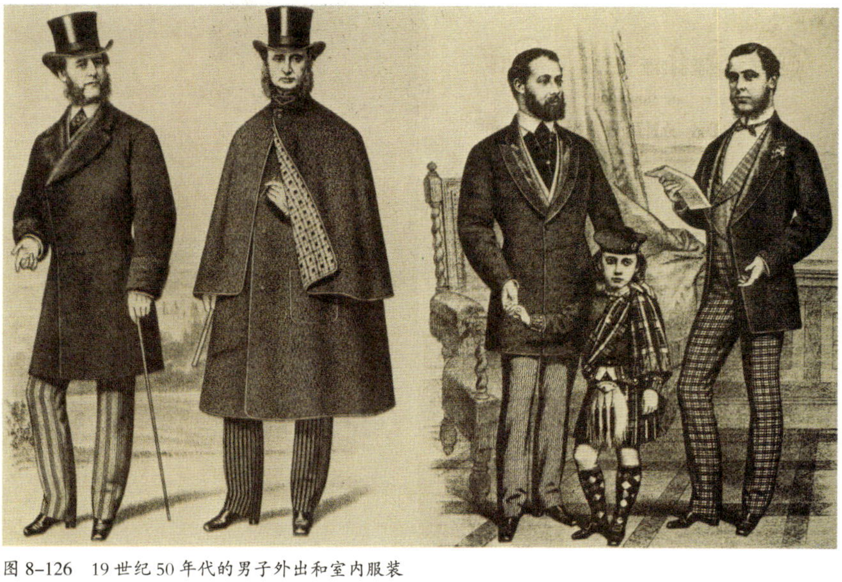

图 8–126　19 世纪 50 年代的男子外出和室内服装

图 8–127　身着户外服的 Albet 王子像

第九章　公元 20—21 世纪

第一节　中国服装

一、民国服饰

清帝退位，民国肇建以后，民国初年实施的新服制和孙中山提出的服装制作四原则，使民众的穿衣戴帽摆脱等级制度和传统的政治伦理的干预，标志着中国古代衣冠体制的解体。

短袄长裙

20 世纪的最初 10 年，中国女性服饰还是以上衣下裙的形式为主。青年妇女往往下穿黑色长裙，上身穿窄而修长的短袄，如三色花缎女短袄（图 9-1），窄身大襟，直袖立领，两侧开衩，面料为湖蓝地花缎，粉红色丝绵衬里。盘纽，领口、斜襟各 1 对，侧门襟 3 对。有的短袄袖子短且肥大，时称“倒大袖”（图 9-2）。这时期还流行一种领高至双耳，遮住面颊的元宝领，被人戏称“朝天马蹄袖”（图 9-3、图 9-4）。虽然领子要高高立起遮住面颊，但小臂却是要露出来，作家张爱玲（图 9-5）称这种式样为：“‘喇叭管袖子’飘飘欲仙，露出一大截玉腕。短袄腰部极为紧小。”民国时期的女裙仍然保留宽大的裙腰，但裙门结构已经消失。辛亥革命之后，西式服饰影响加剧，裙腰系带被更容易穿着的松紧带所代替。在正规场合，裙子内要套长裤和套裤。

图 9-1　三色花缎女短袄

图 9-2　民国时期的倒大袖短袄和长裙

图 9-3　民国元宝领长袄

图 9-4　身穿元宝领长袄的民国女性照片

图 9-5　身穿短袄的张爱玲

民国服制

民国建立以后，缠过又放开的脚多了起来，受西风东渐的影响，喜欢着西式服装的人越来越多。中式衣服，虽然还是两件的袍服，却因为裁剪的关系，变得修长并有线条，极具美感。

民国元年（1912 年），“北洋政府”颁发的服制条例（图 9–6）规定：男子礼服分为“大礼服”“常礼服”两款。常礼服又分为甲、乙两种，大礼服和常礼服的甲种都是西式的。乙种是中式齐领右衽的长袍、短褂，褂是对襟，左右及后开衩。女子礼服一式，是齐领对襟长上衣，下服打裥长裙，俗称百褶裙。

20 世纪 20 年代末，民国政府重新颁布《服制条例》（图 9–7）。这次对国民“礼服”和公务人员制服进行了规定，国民男式礼服为蓝袍和黑褂，女式礼服为蓝袍和蓝衣、黑裙两种款式。男公务员制服为中山装。其形制为立翻领，对襟；前襟五粒扣，代表五权（行政、立法、司法、考试、监察）分立；四个贴袋，表示国之四维（礼、义、廉、耻）；袖口三粒扣，表示三民（民族、民权、民生）主义；后片不破缝，表示国家和平统一之大义。

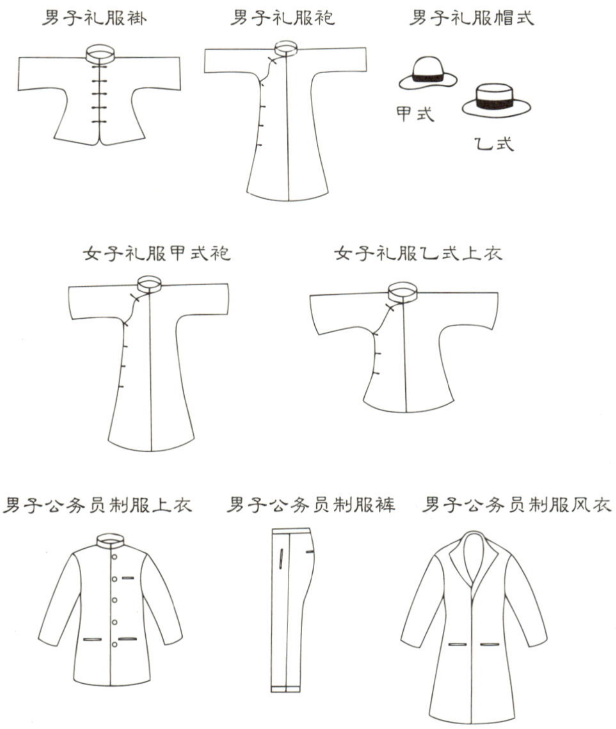

图 9–7　民国时期《服制条例》

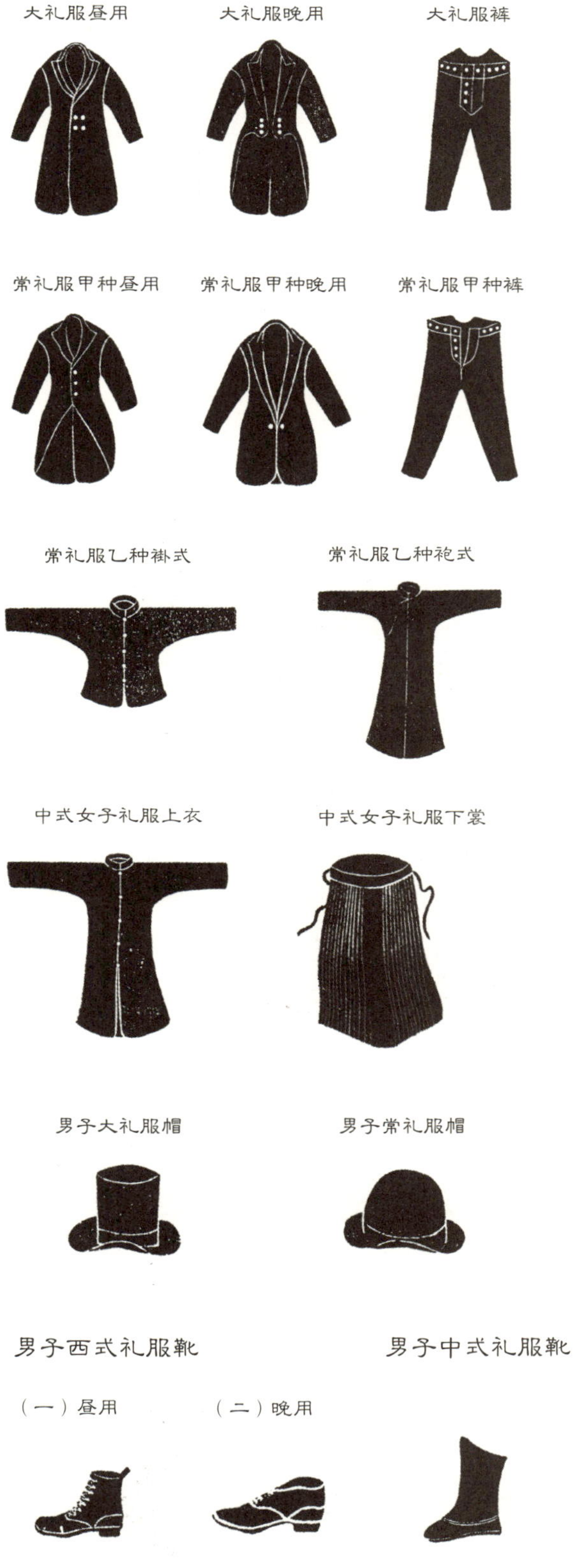

图 9–6　民国元年的《服制条例》

海派旗袍

20世纪20年代至40年代，受西方文化的影响，在我国最早步入大都市行列的上海，曾经出现过时装业蓬勃发展的短暂时期。作为全国服装业的中心，摩登的上海女郎成为全国时尚的领军人物，“中西合璧”的海派服装也形成于该时期。

中国最具民族特色且影响最大的服装莫过于旗袍（图9–8～图9–11）。受西方文化影响，民国时期上海旗袍发生了重大转变。首先，原本宽大的廓型开始变得紧身，能够将东方女性优美的身体曲线表现出来（图9–12）。其次，清代旗袍的封闭性被两侧的高开衩打破。再次，新式旗袍出现了许多局部变化，在领、袖处采用西式服装装饰，如荷叶领、开衩领、西式翻领及荷叶袖、开衩袖，有的下摆坠有荷叶边并做了夸张变形。最后，有些旗袍还大胆使用透明蕾丝材料搭配里面的衬裙。

从服装设计的角度来说，这种旗袍运用了很多设计元素，很有创意，被称为“别裁派”旗袍（图9–13），在当时的“时装片”中盛行一时，而且流行西化服饰的大前提下，“别裁派”旗袍在银幕上频频亮相，成了新颖别致的电影服装。

图9–8　蓝地彩印花罗夹旗袍

图9–9　暗花纹棉麻旗袍

图9–10　湖绿色绸刺绣缘饰旗袍

图9–11　红黄渐变色线钩针编结旗袍

图9–12　1930年旗袍式样

图9–13　“别裁派”旗袍

图 9-14　中山装

图 9-15　列宁装

图 9-16　军便装

图 9-17　日本宽银幕电影《追捕》中的杜丘和真由美

图 9-18　《大西洋底来的人》剧照

二、新中国成立以后服饰

1949 年中华人民共和国的成立，标志着中国服装史步入了一个新时期。至 20 世纪 70 年代，中山装、工装衣裤、列宁装、军便装、方格衬衫和连衣裙等服装式样先后流行。穿草绿色军装，戴草绿色军帽，扎宽皮带，佩戴毛主席像章，挎背草绿色帆布挎包，成为当时的时髦装束。蓝、灰、黑成为大街上的“老三色”，中山装（图 9-14）、列宁装（图 9-15）和军便装（图 9-16）则成为中国人民很长时期的“老三装”。

20 世纪 70 年代末，伴随着改革开放的大门被敞开，人们的审美视野也一并打开了。随着中日电影文化交流的不断深化，一批优秀的日本电影被陆续引进中国。国人在日本宽银幕电影《追捕》中，痴痴地看着冷面硬汉杜丘（图 9-17）和长发美女真由美相拥驰骋在一望无际的草原上，被浑厚男低音主题曲中的“啦呀啦”唱得心驰荡漾。杜丘的风衣、鸭舌帽，后来又加上美国连续剧《大西洋底来的人》（图 9-18）中的蛤蟆镜、喇叭裤，以及大鬓角，成为男孩子的扮酷行头。山口百惠的学生裙、三浦友和的鸡心领毛衫，都被那个年代的观众欣然接受和效仿。1980 年，首部国产爱情片《庐山恋》公映后，连衣裙成为女孩子最钟爱的时装。1985 年，国产影片《红衣少女》在社会上引起轩然大波，其中主人公的大红衬衫，打破了当时中国灰色卡其布服装一统天下的格局。

第二节 西方服装

变革先驱

兴起于18世纪后半叶的英国产业革命、18世纪末的法国资产阶级大革命，加速了西方封建主义身份制度的崩溃，先前繁琐的服饰礼仪也随之消亡解体。现代男装服饰与当代民主社会人人平等的政治体制相适应，突破阶级局限，适用于所有人群。

那些显示性别和物质炫耀的服饰式样被全部抛弃了，欧洲男性的服装风格开始与工业化的节奏相匹配，西方男装趋于简练而具力度：宽平的肩、直线条腰身，方方正正的造型，把现代男性的力量感、效率感描绘得淋漓尽致。此时的西方男性更加注重如何展现自己高贵的生活品位和良好的礼貌教养，男人放弃了对华美的追逐，他们以实用为导向。就服饰对男士所具有的重要性而言，男性尽最大努力追求正确的穿戴，而并非穿着考究或精致。

19世纪西方现代时尚的开拓者乔治·波·布鲁梅尔（Georse Brummell）曾提出，优雅的绅士应该通过服饰细节将自己与出身低的人们区别开来（图9-19）。乔治·波·布鲁梅尔（Georse Brummell）主张，绅士们不应再戴爵位绶带及勋章来表明其贵族的家系，反之，他们应穿戴普通的服装。这些服装应由伦敦萨维尔街(Savile Row)最好的裁缝制成。

20世纪初的巴黎时装进入了由设计师创造流行的新时代。查尔斯·弗雷德里·沃斯（Charles Frederick Worth）是巴黎高级时装业的第一位时装设计师（图9-20）。在时装设计上，沃斯（Worth）摒弃了新洛可可风格的繁缛装束，将女裙的造型变成前平后耸的优雅样式（图9-21）。他的设计风格华丽奢侈，喜欢在衣身装饰精致的褶边、蝴蝶结、花边和垂挂金饰等（图9-22）。沃斯（Worth）在时装界另一项首创是使用时装模特。他也是时装表演的始祖。

图9-19 19世纪西方现代时尚的开拓者 Georse Brummell

图9-20 巴黎时装设计先驱Worth和其1870年的礼服设计稿

图9-21 Worth于1898年设计的丝绸宫廷晚装

图9-22 Worth于1885年设计的结婚礼服

图 9-23　Paul Poiret

图 9-24　Paul Poiret 与模特

图 9-26　Paul Poiret 设计的歌剧外衣

图 9-27　Paul Poiret 设计的作品

图 9-28　20 世纪初期，Paul Poiret 阿拉伯风格时装

图 9-25　Paul Poiret 设计的“孔子袍”

保罗·波烈（Paul Poiret）

保罗·波烈（Paul Poiret）是 20 世纪初巴黎最伟大的时装设计师之一（图 9-23）。他将妇女从紧身胸衣里解放出来（图 9-24）。1910 年，他设计了宽松腰身、膝部以下收紧，具有东方和服风格的霍布尔裙（Bobble Skirt，蹒跚地走路之意）。为步行方便，他在收小的裙摆上做了一个深深的开衩，这是服装史上第一次在女裙上开衩。19 世纪末英法联军和 20 世纪初，欧美游客、商贾在中国的游历、贸易，以及八国联军的劫掠使清代服饰大量流入欧美市场。保罗·波烈（Paul Poiret）也深受东方风影响，推出了“孔子袍”（图 9-25）。其后，保罗·波烈还大胆吸收阿拉伯穆斯林妇女服装的宽松、随和样式（图 9-26 ～图 9-28），引起了法国时装界的轰动。

香奈儿（Chanel）

香奈儿（Chanel，1883—1971 年）开创了现代女性时装（图 9–29），是第一位将男装结构和男装面料、毛针织物用在女装上的设计师。她崇尚的运动和简洁几乎成为一个时代的服装精神（图 9–30），将传统女装的繁文缛节缩减到极限，推出了针织面料的男式女套装、长及腿肚子的裤装、平绒夹克以及长及踝的夜礼服等（图 9–31）。在技术上，她大量借鉴男装的缝制技巧，创造了极尽隐蔽的工艺技术。

图 9–29　Chanel 照片

图 9–30　Chanel 时装画

图 9–31　1921 年法国时尚杂志上的女装

图 9–32　二战时期，欧洲国家和美国参加工厂劳作的女性

图 9–33　20 世纪 20 年代的女装

图 9–34　20 世纪 30 年代的巴黎女子时装

男童风尚

第一次世界大战期间，欧洲男女比例失调，社会劳动力大量缺失，这为女性们走出家庭，进入社会从事生产劳动、参加社会服务提供了机会（图 9–32）。为了工作便利，符合机能性、方便活动的职业服装就成为战争时期女装的发展重点。

这个时期的服装变化需求从以往的晚装礼服转向白天服装。20 世纪 20 年代的女性在外观形象上崇尚帅气、潇洒的气质。无论裙装还是上衣，都有套头装（Jumper Blouse）的趋向。因为采用了套头方式，因此也无须腰带之类的束带，活动和穿脱都自由容易了。由于这种外形很像未成年的少年体形，故称男童式（Boyish，图 9–33）风格。

第一次世界大战后，欧洲女装重新恢复到强调女性的妩媚、娇嫩和雅致的气质上来（图 9–34），流行经过精致剪裁，不放过身体任何细节，但又有节制地突显体形的简洁丝绸长裙。“成熟、妩媚”取代了先前“年轻、稚气、帅气”的形象，体态轮廓也以曲线玲珑有致的“流线型”（Streamline Style）取代以前的“直线型”（Linear Pattern）。烫发技术的出现，也使得卷曲的短发成为流行，与之相配的是斜扁帽。

图 9-35　第二次世界大战期间的西方女性服装

图 9-36　战争期间软木制作的鞋跟

图 9-37　Madame Gres 用红白蓝的法国国旗色设计的时装

图 9-38　Elizabeth Arden 设计了一种穿晚礼服时戴的防毒面具

战时服装

第二次世界大战的爆发引发了物资紧缺，20 世纪 30 年代花花公子和晚会女郎的盛装，被 40 年代轻便、耐用和务实的服装所取代。英、美等国政府开始限制民间纺织品消费，实行节约法令，加之强烈的社会责任和经济紧缩，让人们无暇顾及穿衣打扮。女装也更趋向制服化和功能化（图 9-35）。从 1934 年开始，女装开始变得严谨，加宽了肩部，到 1938 年，为了强调和夸张肩部，上衣服装中开始使用垫肩。受其影响，女性下身的裙子则开始缩短，鞋子则在造型、设计和色彩上更加讲究。因为材料的匮乏，人们开始注重服装材料的品质，那些结实耐用的材料受到重视，尤其是棉麻等与皮肤接触时舒服的材料。此时，人们甚至使用软木制作鞋跟（图 9-36），使用木片制作腰带、手袋等，这样可以节省皮革。

1942 年，法国进入了反法西斯入侵的高潮，法国妇女头戴高而夸张、颜色醒目的帽子，用长发、短裙表达自己的愤怒，帽子和鞋子越来越夸张。在 1943 年德军占领巴黎期间，被誉为“布料雕刻家”的葛莱夫人（Madame Gres）用法国国旗色红白蓝完成了自己复出的第一个系列（图 9-37），她甚至在自己时装屋的一楼窗外挂上了巨大的红色、白色、蓝色三色标识。伊丽莎白・雅顿（Elizabeth Arden）设计了一种穿晚礼服时戴的防毒面具（图 9-38）。

迪奥（Dior）与“新风貌”（New Look）

从1947年“新风貌”（New Look）诞生，到1957年克里斯汀·迪奥（Christian Dior）逝世为止的十年间，高级时装进入了鼎盛时期。

1947年2月12日，法国的天才时装设计师克里斯汀·迪奥（Christian Dior）首次开办了高级时装展，推出名为“新风貌”（New Look）的时装系列，以强调女性隆胸丰臀、纤细腰肢、柔美肩形曲线的设计，打破了战时女装保守古板的线条（图9-39），也改变了战前风靡一时的香奈儿（Chanel）式时装。

图9-39　Dior设计的改变二战后法国时尚风貌的“New Look”

20世纪50年代初，迪奥（Dior）又推出了“垂直造型”及“郁金香造型”。1952年，迪奥时装开始放松腰部曲线，提高裙子下摆。1953年，迪奥（Dior）更是把裙底边提高到离地40厘米（图9-40）。随后几年，迪奥又推出了收减肩宽、增大裙摆的“H型”“Y型”“纺缍型”“A-line型”（图9-41）系列。可以说，20世纪的前半个世纪，高级时装店的设计师们一直是时尚潮流的主宰者。

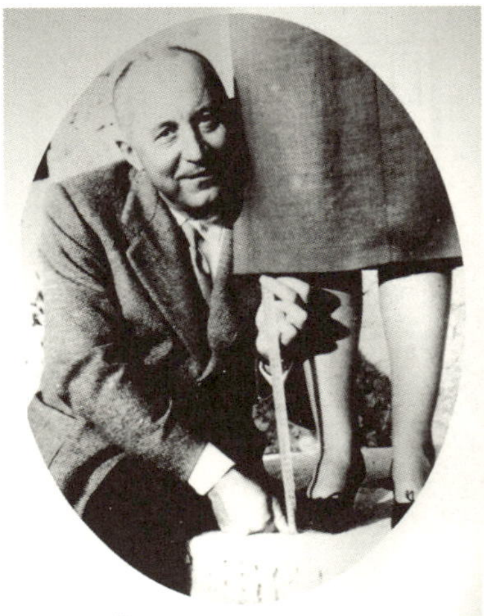

图9-40　时装设计大师Dior

图9-41　Dior的“A-line型”时装

1957年，52岁的克里斯丁·迪奥（Christian Dior）死于心脏病发作。他的去世宣告了高级时装设计的黄金时代结束，使高级时装“型”的时代成为历史。时装的机能性越来越受到重视，这也使面向大众消费群的成衣化生产模式成为可能。随后，巴黎世家（Balenciaga）推出了放松腰身，装饰一个蝴蝶结的“袋形女装”（Sack Dress，图9-42）、强调天真可爱的“娃娃式女装”（Baby Doll，图9-43）；迪奥（Dior）店的继承人伊夫·圣-洛朗（Yves Saint-Laurent）推出了高腰身的“梯形”（Trapeze Line，图9-44）；皮尔·卡丹（Pierre Cardin）推出了高腰身的后背宽松的“斧形”（Serpe）服装，并且裙长缩短到离地50厘米。

20世纪60年代末期，法国社会的不安定，使得高级时装店面临着顾客急剧减少的窘境，而时装店的员工要求加薪的呼声更是让高级时装店面的经营雪上加霜。1970年以后，时装界进入了高级时装与成衣共存的时代。

图9-42　袋形女装

图9-43　娃娃式女装

图9-44　梯形时装

迷你裙（Miniskirt）

20世纪60年代初，伦敦加纳比（London Carnaby Street）附近的女孩们，已开始穿着裙摆至膝部的短裙了。1965年，马丽·奎恩特（Mary Quant，图9–45）在英国白金汉宫隆重推出了刚刚能遮住膝盖的迷你裙（Miniskirt）（图9–46）。随后，安德烈·库雷热（André Courrèges）将裙摆线条再次上移，直到人体膝盖以上，发表了史无前例的“露出膝盖”的迷你装。它被视为“年轻人时装的第一股旋风”的起因，并迅速在欧美风靡。美国全国广播公司（NBC）电台利用刚刚发射成功的卫星，将年轻风貌以最快的速度传播到美国。1966年春夏，纪梵希（Givenchy）和巴尔曼（Balmain）也适应潮流将下摆移至膝盖处，巴黎世家（Balenciaga）和格雷（Gres）夫人则推出膝盖以下的服饰（图9–47、图9–48），香奈儿（Chanel）也推出了紧贴膝盖的“香奈儿长度”，路易·威登（Louis Vuitton）则出人意料地将下摆缩至膝盖上22厘米处，是当时主流服饰品牌最短的迷你装。

图9–45 Mary Quant

图9–46 Mary Quant设计的Miniskirt

图9–47 20世纪60年代露出膝盖的Miniskirt

图9–48 Balenciaga的Miniskirt

图9–49 Punk人群着装

图9–50 Vivienne Westwood

朋克（Punk）

发源于伦敦金斯路郊外的一种新势力朋克音乐（Punk Music）在伦敦街头宣告诞生。朋克（Punk）的兴起和盛行，与20世纪60年代的嬉皮士（Hippes）、摇滚乐队及当时蔑视传统的社会风尚有着千丝万缕的联系。他们以极端的方式追求个性，同时又带有强烈易辨的群体色彩。他们穿着黑色紧身裤、印有寻衅的无政府主义标志的T恤衫、皮夹克和衣服上缀满亮片、大头针、拉链的形象，从伦敦街头迅速复制到欧洲和北美（图9–49）。

朋克族愿意在强烈的摇滚音乐背景下，在各种风格独特的小店中购买价钱便宜，可随时更换与丢弃的服装。“朋克女王”维维恩·韦斯特伍特（Vivienne Westwood，图9–50）在伦敦皇家大道上的时装店，最初取名岩石（Rock），后改为世界尽头（World's End），可谓是名副其实的朋克（Punk）之家。

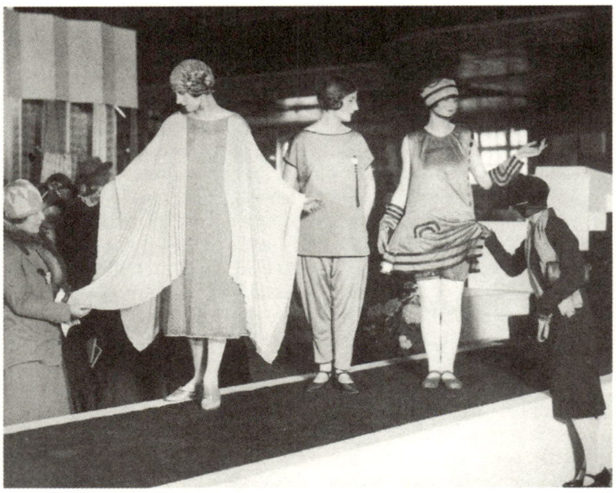

图 9-51　Sorell Fontana 时装秀在 Pitti 宫举办（佛罗伦萨，1953 年）

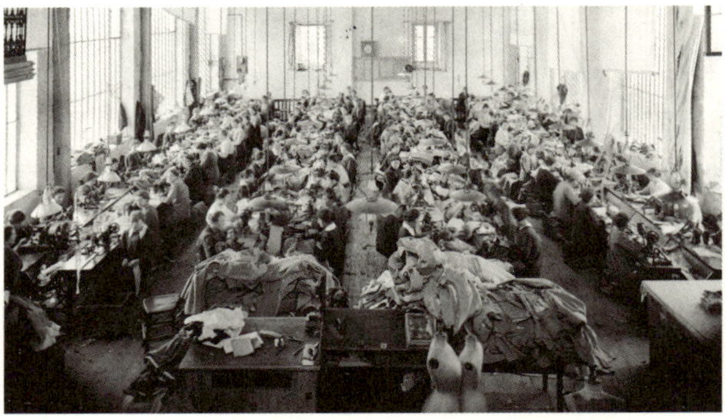

图 9-52　20 世纪中叶的法国服装工厂

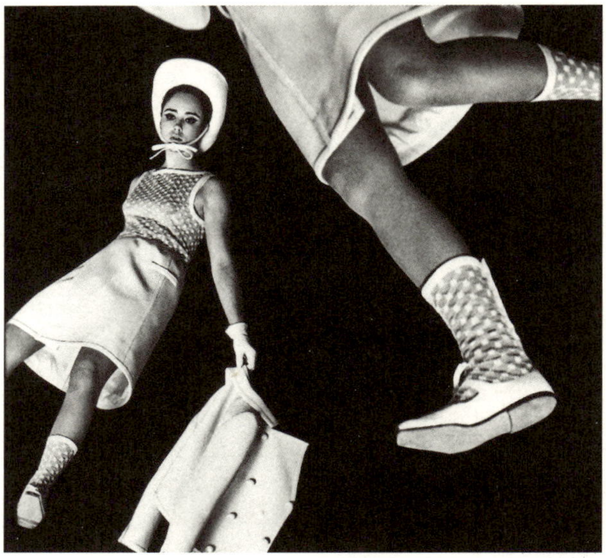

图 9-53　Andre Courreges 的作品

成衣兴起

到了 20 世纪 40 年代，人们对成衣业的印象已开始有所改观。1943 年，公关界的先锋人物埃利诺·兰伯特（Eleanor Lambert）发现美国消费者除了等待着战事的报道之外，还对大洋那边的时尚发展趋势密切关心。于是埃利诺·兰伯特（Eleanor Lambert）将媒体集中到了一起，在纽约召集了一群设计师，举办了世界上第一次时装周（图 9-51）。从此，美国和世界各地的时装掘金者们就定期在纽约举办时装周。

1946 年，人们将“妇女成衣业”（Confection Pour Dames）改名为“女装工业”（Industrie Du Vetement Feminin）。1947 年，在里昂商品交易会上，有 100 个成衣制造商参展。1950 年，法国纺织工业和高级时装业同时跌入困境（图 9-52）。尽管有高级时装协会与纺织品补助政策的大力扶持，高级时装业还是不可阻止地走向衰落。1952 年有 60 家高级时装公司，到 1958 年时仅剩下了 36 家。一些人为了挽救颓势，开始考虑如何将高级时装与成衣联合。

1959 年，德国成衣业协会与法国高级时装协会签订了一个协议，即每年两次由包括巴尔曼（Balmain）、迪奥（Dior）、郎万（Lanvin）等在内的 15 名法国高级时装设计师，每人带 60 至 70 件款式的衣服到德国参展，使德国的成衣制造业能够及时了解服装最新流行趋势，从而降低批量生产的风险性。而法国的高级时装设计师们也能从展会的门票收入中获得丰厚的利润。在 1958 年至 1963 年的这段时间，美国和一些欧洲国家的成衣业正在势不可挡地发展壮大起来，成衣制造商从巴黎的高级时装大师们手中购买新的款式，再回到本国进行批量生产。20 世纪 60 年代初，每季都有满满一架飞机的新款式衣裙和样板从法国运往纽约。

经过多年的努力，人类终于在 1969 年登上了月球，这促成了太空风貌和现代主义风貌的流行。曾经当过飞行员的时装设计师安德烈·库雷热（Andre Courreges）在 1965 年春夏季推出了包括帽子、眼镜、手套在内的宇航员式样的时装系列（图 9-53）。太空时代的无性别区分的特点成了时装界一个大胆的尝试，喇叭裤与靴子代替了迷你裙与长统靴或低跟女鞋。此时，皮尔·卡丹（Pierre Cardin）也推出了很多太空风貌的时装作品。

图 9-54 20 世纪 80 年代的 Yuppie Look

物质年代与雅皮士（Yuppies）

20 世纪 80 年代是一个享受的年代、丰裕的年代。物质主义成为生活的中心，追求物质享受成为这个时期的中心。冷战结束、苏联的戈尔巴乔夫改革、波兰的团结工会、中国的改革开放政策都显示出那是一个宽松的时代，整个世界的中心是政治稳定、经济增长和经济扩张。“Work hard and play hard”成为人们的生活准则。

在时装方面，穿着讲究，年轻人反叛的时装风格已经被归入主流时尚。这个时期，曾经激愤的青年人已开始成为社会的主流，嬉皮士（Hippies）风格演变成雅皮士风貌（Yuppie Look）（图 9-54）。20 年前的以精神至上、意识形态为主导的文化被物质社会的感官享受所取代。他们的着装、消费行为及生活方式等带有较明显的群体特征，但他们并无明确的组织性。

雅皮士（Yuppies）有着较优越的社会背景，如较高的社会地位、丰厚的薪水等。他们不一定年轻，但他们对奢华物品、高级享受的追求热情十足。雅皮士（Yuppies）衣着讲究，修饰入时，处处透露出他们所拥有的良好的生活状态。

图 9-55 1980 年的时尚风貌

在价格不菲的西服袖肘处镶拼一块椭圆形皮质补丁曾是雅皮士或有雅皮倾向的人士所喜爱的。你似乎很难从这里找到任何美感的依据，但毫无疑问，它容易让人想起高级写字楼。

雅皮士着迷于法拉利（Ferrari）跑车、劳力士（Rolex）手表、古驰（Gucci）饰品、范思哲（Versace）时装的主要原因之一是他们对这些高档物品让一般人不敢问津的价格所代表的质量和信誉深信不疑。

20 世纪 80 年代的职业女性，要在各个领域与男性一争长短，所以采取了非常咄咄逼人的态度。如剪裁精致、宽垫肩的上衣，短而紧身的裙子和讲究的衬衣。宽而棱角分明的垫肩是从男士西服中借鉴过来的，直而硬的廓型传达出权威的感觉（图 9-55）。

后 记

本人以为，在研究服装史、撰写服装史的专著、教材时，有两种情况是比较难做好的：一种情况是对服装史进行深入刨析、系统阐述，将每一个细节都研究到且阐述透彻；另一种情况是把服装史的书写薄了，写简单了，即建立大框架、大关系，把基本概念讲清楚了，将不同服饰等级、不同服饰制度之间的关系讲清楚。总之，越简单清晰越好。当然，能做到简单也很不容易，不仅需要做到将服装史的内容了然于胸，还要具备能够将错综复杂的内容化繁为简的视野。

记得当年在中国服装史泰斗清华大学美术学院黄能馥先生家里，听黄先生讲课，他说：“我们在写服装史著作和教材的时候，要想办法深入浅出，把最深奥的道理，用最简单的话讲出来，让读者清晰准确地了解中国服装史的内容。”虽然已经过去十几年，但黄先生当年的教诲让我铭记于心。此外，我在清华大学实际教学过程中也发现了这个问题。有的时候概念太多、内容过于深奥，对于初学者来讲是有一定困难的，不能快速、清晰地建立“服装史”的知识体系和内容框架。

《中国服装史（简明版）》《西方服装史（简明版）》《中外服装史（简明版）》三本书，就是力求用最简单的语言将服装史的内容阐述清晰，使读者在最短的时间内了解服装史的概貌。采取简单、明了、准确的方式让读者快速地学习，并且建立起这些基本的知识体系，然后在此基础上，进行更深一步的研究，我觉得这是学习服装史的一个比较好的方法。在这种理念的指导下，我撰写完成了这套服装史简明版系列，希望能对读者有所帮助。

2021年5月18日写于清华大学